Advances in

Plant Pathology

Volume 1

Advances in Plant Pathology

Volume 1

edited by

D.S. Ingram

The Botany School
Downing Street
Cambridge
England

and

P.H. Williams

Department of Plant Pathology
The University of Wisconsin
Madison, Wisconsin
USA

1982

ACADEMIC PRESS

A Subsidiary of Harcourt Brace Jovanovich, Publishers

London New York
Paris San Diego San Francisco
São Paulo Sydney Tokyo Toronto

ACADEMIC PRESS INC. (LONDON) LTD.
24/28 Oval Road,
London NW1 7DX

U.S. Edition published by
ACADEMIC PRESS INC.
111 Fifth Avenue
New York, New York 10003

British Library Cataloguing in Publication Data
Advances in plant pathology.—Vol. 1
1. Plant diseases
581.2 SB731

ISBN 0-12-033701-0

Printed in Great Britain at the Alden Press, Oxford

Contributors

J.J. BUJARSKI *Department of Horticulture, University of Wisconsin, Madison, Wisconsin*

J.M. DALY *Department of Agricultural Biochemistry, Institute of Agriculture and Natural Resources, University of Nebraska, Lincoln, Nebraska*

J.E. ELLISTON *The Connecticut Agricultural Experiment Station, New Haven, Connecticut*

T.C. HALL *Agrigenetics Advanced Research Laboratory, Madison, Wisconsin*

N.T. KEEN *Department of Plant Pathology, University of California, Riverside, California*

H.W. KNOCHE *Department of Agricultural Biochemistry, Institute of Agriculture and Natural Resources, University of Nebraska, Lincoln, Nebraska*

D.J. MERLO *Division of Biological Sciences, University of Missouri, Columbia, Missouri*

W.A. MILLER *Department of Horticulture, University of Wisconsin, Madison, Wisconsin*

Preface

Advances in Plant Pathology is a new series which will publish extended review articles on topics of current or future interest and importance in plant pathology. We intend that these articles shall be idea-orientated, rather than mere reviews of the literature, and will be controversial, where appropriate. They will also be well written and presented, and aimed at a wide audience including undergraduate and postgraduate students, research scientists and college teachers. Authors, who will include both established scientists and those at the start of their careers, will be given as much space as they require to develop their ideas fully, without undue restriction as to the length, style and format of their contributions. In some cases, articles will be of monograph length, if the subject covered is of sufficient breadth and importance.

Most volumes in the series will not have any particular theme, but will contain a mixture of articles drawn from a range of subject areas within the field of plant pathology. Volume 1 has a distinct genetical/physiological flavour, but this is entirely fortuitous, reflecting the availability of completed articles rather than the deliberate intentions of the editors. Future volumes, in addition to drawing articles from the subject areas dealt with in Volume 1, will carry reviews on a wide variety of plant pathological topics including bacteriology, virology, genetics, physiology/biochemistry, ecology, epidemiology, mathematical modelling, control, breeding for resistance, and descriptive pathology. Also, from time to time, articles will be included which deal with topics that are not strictly plant pathological, but which the editors consider to be highly relevant to plant pathology itself.

Despite the wide-ranging character of the majority of volumes, some will deal with a single important subject in depth, either as a series of short articles by several different authors or, at the other extreme, as a complete work by a single author. Finally, where appropriate, a single volume may be devoted to the proceedings of a symposium or conference providing that we consider the contributions to be of sufficient quality and wide interest.

It is through its flexibility of style, format and content, its emphasis on ideas and interpretations rather than mere reviewing of the literature, and the breadth of its target audience that *Advances in Plant Pathology* will differ from other review series. We are anxious to serve the needs and interests of all plant pathologists, and would therefore *welcome* comments, suggestions and ideas for future volumes in the series. We also wish to make it clear that although many of the articles in the series will be specially commissioned, we are also willing to consider for publication articles submitted to us direct, without prior invitation. Any such articles should be sent to Dr D.S. Ingram, *The Botany School, Downing Street, Cambridge CB2 3EA, England.*

Cambridge and Madison, 1982

D.S. Ingram
P.H. Williams

Contents

Contributors . . . . *v*

Preface . . . . *vi*

1. Hypovirulence
J.E. Elliston 1

2. Specific Recognition in Gene-for-Gene Host–Parasite Systems
N.T. Keen 35

3. The Chemistry and Biology of Pathotoxins Exhibiting Host-Selectivity
J.M. Daly and H.W. Knoche 83

4. Crown Gall: A Multipotential Disease
D.J. Merlo 139

5. Enzymes Involved in the Replication of Plant Viral RNAs
T.C. Hall, W.A. Miller and J.J. Bujarski 179

Index 213

1

HYPOVIRULENCE

J. E. Elliston

The Connecticut Agricultural Experiment Station

I. Introduction
II. Concepts of Virulence and Hypovirulence
III. The *Castanea-Endothia* Pathosystem
   A. Chestnut Blight in North America
   B. Chestnut Blight in Europe and the Discovery of Cytoplasmic Hypovirulence
   C. Chestnut Blight in Asia
   D. North American Cytoplasmic Hypovirulent Strains
   E. The Probable Nature of Cytoplasmic Hypovirulence Agents in *E. parasitica*
   F. Hypovirulence and Natural Biological Control of *E. parasitica* in Europe: A Biological Puzzle
   G. Possible Origin of Hypovirulence in *Endothia*
IV. Hypovirulence in Other Pathosystems
V. Concluding Remarks
References

I. INTRODUCTION

Whenever the functioning of tissues essential to a plant is adversely affected, by whatever means, non-specific processes are set in motion that serve to protect it from further damage and to restore or replace these tissues and functions (Mullick, 1977). These processes are the plant's active defence mechanisms. The plant developed them in response to organisms that evolved with it, and to changes in its physical environment. In general, they protect it adequately from serious disease and premature death.

Serious disease develops when a pathogen disrupts, suppresses or avoids triggering these defence processes. In a balanced pathosystem, serious disease occurs in a small fraction of the suscept population and serves to eliminate the least fit individuals. Failures of defence that threaten the survival of a species probably are rare in nature. Man, however, has initiated such failures by carrying pathogens across natural barriers that previously had isolated them from related, less resistant suscepts. Four notable examples are blight of North American and European chestnut species, incited by *Endothia parasitica*, blister rust of five-needled pines, incited by *Cronartium ribicola*, Dutch elm disease of North American and European elm species, incited by *Ceratocystis ulmi*, and Scleroderris canker of many North American gymnosperms, incited by *Gremmeniella abietina*.

Plant pathologists have attempted to deal with such failures in defence by hybridizing the suscept with its more resistant relatives, adjusting cultural practices to favour the suscept, triggering latent resistance mechanisms in the suscept, eradicating the pathogen or its alternate host(s), and creating external or internal toxic or inhibitory barriers to the pathogen.

Another approach is to afflict the pathogen with a pathogenic hyperparasite. The object of using a hyperparasite is to reduce the pathogen's ability to incite disease by disrupting its mechanisms of pathogenesis, reproduction, or both. If the hyperparasite becomes established, a balance between suscept and pathogen may develop; the defence mechanisms that the suscept evolved to cope with its accustomed parasites and pathogens may allow it to cope adequately with the weakened alien pathogen.

Hybridization between the alien pathogen and less virulent native relatives could also help to balance the pathosystem, in the same way as hybridization between related resistant and susceptible suscept species. Hypovirulence, a phenomenon first associated with the chestnut blight fungus, could be a consequence of hyperparasitism and pathogen hybridization.

II. CONCEPTS OF VIRULENCE AND HYPOVIRULENCE

Biologists have used the term "virulence" to refer to the disease-producing capacity of a pathogen (Steen, 1971). I consider virulence to be synonymous with pathogenic fitness *sensu* Nelson (1979), i.e. fitness to infect, colonize, kill and reproduce in suscept tissue. It might be viewed quantitatively as a continum in both the pathogenicity and reproductive capacity of a pathogen.

Hypovirulence, which literally means subnormal virulence, includes all abnormal states in which pathogenic fitness is reduced. It can have many causes. Distinctions between types of hypovirulence are needed and might best be founded on whether the genetic information that determines them is nuclear or cytoplasmic. *Nuclear hypovirulence* is due to nuclear genetic determinants and might be caused by mutant nuclear genes, heterokaryons, hybrid nuclei or extrachromosomal genetic determinants that reside in the nuclei. *Cytoplasmic hypovirulence* (CH) may be conditioned by agents such as viruses and virus-like agents, plasmids and organelles that carry genetic elements, such as mitochondria. Since many of the CH agents may themselves be considered as hyperparasities, CH caused by these agents may be viewed as a consequence of hyperparasitism. Synonyms that have been used previously for CH include exclusive hypovirulence (Grente and Sauret, 1969a), contagious hypovirulence (Grente, 1975), transmissible hypovirulence (Van Alfen, *et al.*, 1975) and hypovirulence (Jaynes *et al.*, 1976; Day *et al.*, 1977).

This classification of hypovirulence may prove inadequate as knowledge of the phenomenon increases. It is likely that genetic determinants conferring hypovirulence may reside simultaneously in the nuclei and cytoplasm of a pathogen. It is also likely that cases will be found in which nuclear and cytoplasmic determinants operate together to reduce the pathogenic fitness of an organism.

III. THE CASTANEA–ENDOTHIA PATHOSYSTEM

The *Castanea–Endothia* pathosystem involves three distinct sets of genotypes: those within the genus *Castanea*, those within the fungus *Endothia parasitica* and, perhaps, its close relatives in the genus *Endothia*, and those within the "species" of CH agents that affect *E. parasitica* and its relatives. What follows is an attempt to summarize what is known about this pathosystem, and to suggest ideas about how hypovirulence might have arisen, and how it might operate to restore biological balance to this system.

A. Chestnut Blight in North America

The chestnut blight fungus, *Endothia parasitica* (Murr.) P. and H. And., an ascomycete, was first noticed during the summer of 1904 causing a serious canker disease on American chestnut trees, *Castanea dentata* (Marsh.) Borkh., in the grounds of the New York Zoological Park in New York City (Murrill, 1906a, b). The fungus apparently had been introduced some years earlier, perhaps in the 1890s, because immense numbers of trees in the New York City area were already dead or dying when Murrill's attention was brought to the problem.

Fig. 1. Blighted American Chestnut. (a) girdling canker subtended by epicormic sprouts; (b) mycelial fans of *Endothia parasitica* in infected bark; (c) erumpent stromata containing perithecial beaks (black dots).

Cankers occur anywhere on the branches and trunk (Fig. 1a). The fungus appears to require a wound to gain entrance, but once infection is established, its mycelial fans (Fig. 1b) colonize the bark tissue very rapidly, encircling and girdling the branch or trunk in one or more

Fig. 2. Ninety-year-old stand of American chestnut before the chestnut blight epidemic.

seasons, killing the cambium, and preventing movement of water and metabolites to and from parts of the tree distal to the canker. Abundant orange-brown stromata erupt through the surface of infected bark and, under moist conditions, release twisted spore tendrils composed of millions of minute, single-celled conidia embedded in a water-soluble, gelatinous matrix. These spores are adapted for spread by rain and perhaps by insects, birds, and small mammals. Perithecia usually develop abundantly in the stromata (Fig. 1c), and produce ascospores that are released after rains and are adapted for spread by wind.

Murrill recognized the unparalleled virulence of this fungus to the American chestnut as well as to the native chinkapin, *C. pumilla*, a close relative (Murrill, 1908). He quickly concluded that preventative measures and sanitation were useless against it, and predicted that the native chestnut species would become extinct if the fungus continued to spread unchecked.

The natural range of *C. dentata* extended from southwestern Maine south to Georgia and west to eastern Michigan, Ohio, Indiana, Kentucky, Tennessee and Mississippi. The tree had a nearly continuous

Fig. 3. Decaying stump (left), log (right) and current root sprout (centre) of American chestnut tree killed by *Endothia parasitica* early in the century.

distribution and was one of the most abundant in the eastern deciduous forest (Fig. 2). The fungus spread very rapidly through the northern part of the range (Clinton, 1913; Pennsylvania Chestnut Tree Blight Commission, 1915). The massive effort made in Pennsylvania to slow or stop the spread of the disease, by creating chestnut-free zones in advance of the blight and eradicating spot infections as they were detected, was given up as useless two years after it began. Clinton (1924) noted that the fungus spread more slowly in the southern part of the tree's range but provided no explanation for this. By 1950, *E. parasitica* had infected nearly every chestnut tree within the range and had killed most to the root collar.

Today, sprouts from the root systems of blighted trees are all that remain of *C. dentata* within its natural range (Fig. 3). In some areas, sprouts are still abundant and occasionally attain diameters greater than 25 cm before being killed back to the roots by blight. Because *C. dentata* is self-sterile and pollination is rare now, vegetative reproduction perpetuates the species in the forests.

The high level of virulence displayed by *E. parasitica* in *C. dentata*, its

Fig. 4. Remnant European chestnut tree and root sprouts on an estate at Bucalla, near Genoa, forty years after the chestnut blight fungus was discovered there. Nearby sprouts contained abnormal cankers.

abundant production of inoculum adapted for spread by animals, insects, rain and wind, the abundance and nearly continuous distribution of the suscept, and favourable environmental conditions combined to produce an epidemic more destructive than any known before in North America.

B. Chestnut Blight in Europe and the Discovery of Cytoplasmic Hypovirulence

Endothia parasitica was first found in Europe in 1938 in a grove of European chestnut on an estate near Genoa in northern Italy (Fig. 4) (Biraghi, 1946). The fungus was well established and inflicting serious damage when discovered, suggesting that it had been introduced some years earlier (Woodruff, 1946). Its nearly simultaneous discovery near

Genoa in the northwest, Udine in the northeast, and Avellino in the south (Biraghi, 1950a) may be more than coincidental. During the 1920s, experimental nurseries of Japanese chestnut were established in many areas where evidence of early blight infections has been found (Bagnaresi, 1956). These trees may have been the source, and initial infections could have occurred up to 20 years before they were recognized. Woodruff (1946) suggests that these early infections did not spread, and the epidemic originated from subsequent multiple introductions through the widely separated seaports of Genoa, Venice and Naples. Vigorous attempts to contain the new outbreaks by sanitation and quarantine failed in Italy as they had earlier in the USA (Pavari, 1949). Within approximately 25 years of its discovery, blight spread to all major chestnut growing areas of Italy (Mittempergher, 1979).

In 1950, Biraghi (1950b) noticed several chestnut coppices that appeared to be unusually healthy in an area where blight had been severe previously. They were located near the chestnut nursery at Masone near Genoa and had been the site of a vigorous but unsuccessful two-year sanitation experiment 11 years earlier. At the time this early experiment was conducted, living stump sprouts more than four or five years old were rare because rates of infection and mortality were high. Biraghi returned to this site during the summer of 1951 (Biraghi, 1953a, b), and carefully examined the clumps of stump sprouts that comprised one of the coppices. Nearly all sprouts contained at least one infection. On sprouts from some stumps he found only sunken, lethal blight cankers subtended by numerous epicormic sprouts, the typical symptoms of blight. Sprouts from other stumps contained mixtures of normal cankers and unusual, abnormal cankers (Fig. 5). Affected bark in the abnormal cankers was somewhat swollen, and sprouting from bark below the infections either was absent or greatly reduced. Many of the stump sprouts with abnormal cankers showed no sign of debilitation. To Biraghi, the diverse sizes and appearances of these abnormal infections reflected different stages of development. The fungus was confined to the outer layers of the bark, where it occurred as mycelial fans between the strata (Fig. 5d). On many sprouts it had colonized large areas, but the bark tissues and cambium beneath the infections were alive (Fig. 5c). The bark surface was cracked lengthwise and roughened in younger infections (Fig. 5a), while in older infections it had become blackened and broken into smaller scales (Fig. 5b). Other abnormal cankers consisted of exposed sapwood bordered by vigorous callus, with superficial infections radiating from the margins of the openings. Residues of typical *E. parasitica* mycelial fans occurred on the exposed wood and in fragments of dead bark adhering to the margins of the

Fig. 5. Cankers found on European chestnut in Italy in 1978 resembling those described by Biraghi in 1953: (a) young, abnormal, girdling canker with longitudinally split bark; (b) older infection with bark split into smaller scales; (c) old abnormal canker with bark split into small, black, flaky scales. Note live sapwood and inner bark under the superficially infected bark; (d) mycelial fans of *Endothia parasitica* between outer layers of bark in an abnormal canker.

openings. The host had responded vigorously to the fungus, preventing further deep penetration and its usual lethal consequences.

Biraghi proposed that either the sprouts had acquired resistance to the pathogen, or the fungus that produced abnormal infections had lost some of its virulence. He favoured the induced resistance hypothesis (Biraghi, 1953a, b), mainly because abnormal cankers occurred only in coppices with long histories of blight, and the cankers resembled those produced on the more resistant Japanese chestnut. Apparently he was so convinced that he did not vigorously explore the other explanation. If he had, he might still have come to his original conclusion because naturally occurring abnormal cankers on *C. sativa* often contain mixtures of normal and hypovirulent forms of *E. parasitica* (Grente and Sauret, 1969a; Bonifacio and Turchetti, 1973; Elliston, unpublished), and the hypovirulent forms could easily be discarded as contaminants because they often do not resemble typical isolates of *E. parasitica* (Fig. 6a, b, c, d). In any event, his conclusion stood for the 14 years that followed.

In 1964, French scientists sought alternatives to the induced resistance hypothesis to explain the recession of serious blight and regrowth of chestnut in Italy. With the discovery of *E. parasitica* in southern France in 1956, and the threat it posed to the chestnut industry there, they hoped that the basis for the recovery could be determined and this knowledge could be applied to curtail blight epidemics in France and elsewhere. They visited recovering chestnut groves along the slopes of the Italian Alps in the Lombardy region and examined and sampled the cankers (Grente, 1965). Grente and Sauret (1969a) found that many isolates from abnormal cankers did not resemble normal *E. parasitica*. Abnormal isolates typically produced white rather than orange colonies in culture (Fig. 6b, c), sporulated less rapidly and abundantly, and were non-pathogenic or far less pathogenic to *C. sativa* than normal isolates. The identity of the strains was established when some of their conidia and hyphal tips produced normal, wild type isolates of *E. parasitica* (Fig. 6a). The white isolates retarded or prevented canker development when inoculated into chestnut bark together with related normal isolates. White isolates introduced into existing cankers caused by related normal strains arrested canker development and permitted wounds to begin to close. With time, white forms dominated these cankers and the tree shed the infected bark. Grente and Sauret (1969a) coined the term exclusive hypovirulence to describe this behaviour: hypovirulent strains became dominant over normal strains and infected bark was excluded from living tissues of the tree.

With auxotrophic strains they found that factors conferring exclusive

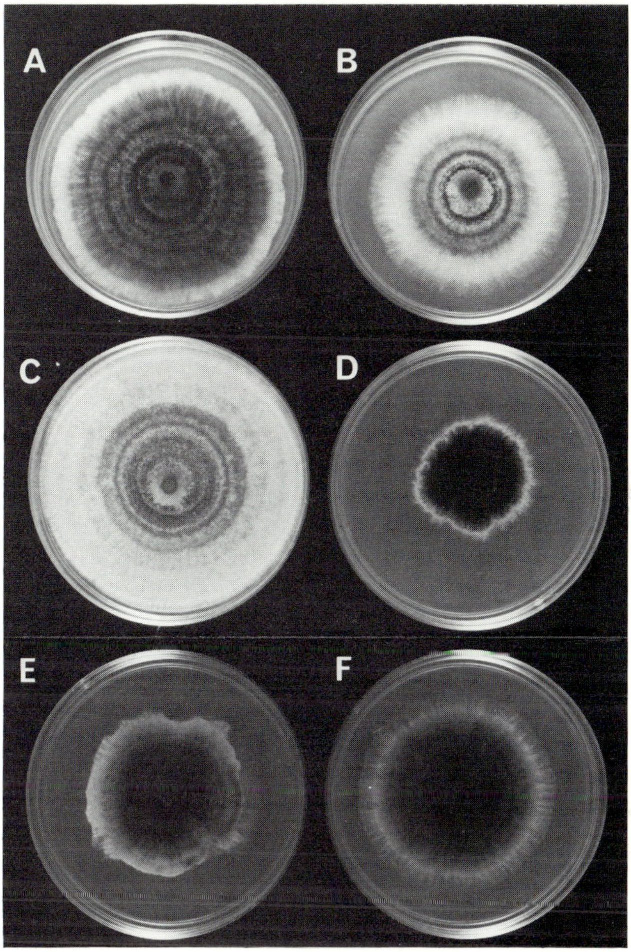

Fig. 6. Normal and abnormal isolates of the chestnut blight fungus: (a) stable, normal isolate; (b) unstable, "white", slow-growing isolate; (c) unstable, "white", fast-growing isolate; (d) stable, dark orange-brown, pigmented isolate; (e) unstable, pigmented isolate with slow, erratic growth; (f) unstable, pigmented isolate with intermediate growth rate. Isolates a–d were obtained from Italy and e and f from North America. All cultures are seven days old.

hypovirulence were cytoplasmic (Berthelay-Sauret, 1973). Thus, some 18 years after abnormal cankers were discovered in Italy, it became clear that their abnormality was due more to reduced virulence in the fungus than to increased resistance in the tree, and that this loss was associated with the presence of infectious cytoplasmic agents.

Subsequently, Grente and Berthelay-Sauret (1979) developed a biocontrol programme in France using CH strains. Their procedure was to isolate the fungus from the grove to be treated, determine which CH strain could transmit CH agents into the strain, then treat 5 to 20 cankers per hectare with the chosen CH strain. They reported that CH usually spread from the treated cankers at about 1 to 2 m/year, and in about 10 years, no new blight infections could be found.

In Italy, the incidence of lethal chestnut blight infections has been declining for more than 25 years, and abnormal cankers have become common (Turchetti, 1979). Italian scientists have isolated white hypovirulent strains from abnormal cankers in many areas of Italy where recovery is occurring (Bonifacio and Turchetti, 1973). This supports the conclusion that CH agents have a role in this recovery phenomenon.

C. Chestnut Blight in Asia

Endothia parasitica apparently is indigenous to Asia. Meyer found it on *C. mollissima* in widely separated regions of China, but the disease differed in severity from place to place (Fairchild, 1913; Shear and Stevens, 1913a; Shear *et al.*, 1917). The chinkapins of China were immune. Meyer also found the fungus causing cankers on *C. crenata* in Japan, and reported the Japanese chestnut to be more resistant than the Chinese chestnut (Shear and Stevens, 1916; Shear *et al.*, 1917). Kajiura (1955) noted that the fungus had been known in Japan for many years and had a wide distribution, and that it had not been studied because it did not cause serious damage. He also noted that selections of Japanese chestnut differ in resistance to the fungus.

Fossil evidence indicates that *Castanea* has had a long history in China (Hu and Chaney, 1940) and in Japan (Eynard *et al.*, 1966). Whether *E. parasitica* evolved in China, Japan, or both is unknown. It is not known if CH agents occur in the Asian populations of *E. parasitica*.

Isolates of *E. parasitica* from China, Japan, and eastern North America did not differ in cultural characteristics or virulence to *C. dentata* (Shear and Stevens, 1913b, 1916; Shear *et al.*, 1917).

D. North American Cytoplasmic Hypovirulent Strains

The first American CH strain of *E. parasitica* (Fig. 6e) was isolated from an abnormal canker on an American chestnut sprout in western Michigan (Elliston *et al.*, 1977), beyond the natural range of *C. dentata*. Except for its orange-brown pigmentation, it did not resemble wild type

Fig. 7. Canker on an American chestnut tree controlled by CH agents from the American strain shown in Fig. 6e.

strains of *E. parasitica* or any of the European CH strains. Although it is non-pathogenic in American Chestnut, a small fraction of its conidia develop into typical wild type strains that have normal virulence. Co-inoculation of this strain and related normal strains in *C. dentata* prevents or curtails canker development and converts the normal strains to the abnormal type (Fig. 7). Three similar CH strains were isolated from abnormal cankers on an American chestnut tree in Tennessee (Elliston *et al.*, 1979). Other presumed CH strains were isolated from persistent cankers on *C. dentata* in Michigan and several other locations within the natural range of the tree (Fig. 6f) (Day *et al.*, 1977; Jaynes and Elliston, unpublished). Like European strains, their cultural characteristics and biological behaviours are diverse, but none resembles the European CH strains. The unusual persistence of some large, chronically infected American chestnut trees (Fig. 8) may be due to CH rather than to unusual resistance.

E. The Probable Nature of CH Agents in *Endothia parasitica*

Cytoplasmic hypovirulent strains of *E. parasitica* consistently contain double-stranded ribonucleic acid (dsRNA) (Day *et al.*, 1977; Day and Dodds, 1979; Dodds, 1980a; Dodds and Elliston, 1978), genetic material typical of most fungal viruses (Lemke and Nash, 1974). Normal

Fig. 8. Large, chronically-infected American chestnut tree in Madison, Virginia. CH strains have been isolated from this tree (Courtesy R. A. Jaynes).

wild-type strains and normal strains derived from CH strains lack detectable quantities of dsRNA. Although a definite cause and effect relationship between presence of dsRNA and CH has not yet been proven, since Koch's postulates have not yet been satisfied, it is reasonable to assume that virus-like agents are responsible for this condition.

The presumed viruses or virus-like agents are unusual. The molecular weights estimated for the major classes of dsRNA extracted from CH strains range between 4·0 and $7{\cdot}0 \times 10^6$ daltons, values that are significantly higher than those of nearly all previously known fungal virus-like agents. The dsRNA patterns from European and American strains are distinct (Dodds, 1980a).

Although few strains have been examined thoroughly, conventional techniques for detecting virus-like particles in cytoplasmic hypovirulent strains have not been successful (Moffitt and Lister, 1975; Dodds, 1979). The dsRNA in European strains appears to be contained in membrane-bound, club-shaped particles (Dodds, 1980b), resembling those associated with a severe disease of cultivated mushroom (Lesemann and Koenig, 1977). Particles of this kind have not been isolated from American strains using procedures that yielded them from European strains (Dodds, 1979). These observations, together with the distinct abnormalities conferred, suggest that different CH agents occur in European and American CH strains.

Unlike most known fungal virus-like agents, these agents significantly alter the properties of their host. Thus, they may constitute a new class of virus-like agents (Dodds, 1980b).

F. Hypovirulence and Natural Biological Control of *Endothia parasitica* in Europe: A Biological Puzzle

Biologists who are aware of the hazards of chemicals for plant pest control (Smith and Pimental, 1978) recognize the merits of biological agents such as those associated with CH in *E. parasitica*. Herbivorous insect pests, particularly introduced species, have been controlled effectively with parasites (Woods, 1974). So far, attempts to use hyperparasites to control plant pathogens have been few and largely unsuccessful (Kuhlman, 1980).

In Italy, CH agents appear to have brought *E. parasitica* under control naturally, but how this has come about is unknown. Present knowledge and possible mechanisms for the phenomenon are discussed below.

1. Diversity of CH Strains

Initially, all CH strains were assumed to be white in culture and weakly pathogenic or non-pathogenic. This view emerged from extensive studies of two CH strains and their derivatives (Van Alfen *et al.*, 1975, 1978; Jaynes *et al.*, 1976). Subsequent studies of a more representative collection of CH strains revealed a wide range of abnormalities in morphology (Fig. 6), pathogenicity, growth rate, pigmentation and reproductive capacity (Elliston, 1977, 1979). This diversity could be due to diversity among agents that confer CH, different effects of individual CH agents in different genotypes of the fungus, or additive, synergistic, or other effects of simultaneous infection with more than one CH agent.

Individual cankers often yield mixtures of hypovirulent strains and normal strains (Grente and Sauret, 1969a; Bonifacio and Turchetti, 1973; Elliston, unpublished). Normal mycelium could occur as pockets uninvaded by the CH agents or as pycnidia or perithecia formed before CH agents entered the thallus. Coexistence of biologically distinct hypovirulent strains in the same canker has several possible explanations. Individual cankers might be produced by several genetically distinct strains of the fungus operating together, and some of these strains might be more resistant to infection by CH agents than others or react differently to their presence. Another possibility is that the population of a CH agent might differ from one region to another within a thallus, and the degree of debilitation might be directly correlated with the concentration of the hypovirulence agent. A third possibility is that different CH agents might be introduced into a canker at different times, yielding different single infections in some portions of the thallus, multiple infections in others, and no infection in still others. Regions of the thallus infected with single CH agents might be less debilitated than regions containing several different agents. Finally, mixtures of CH agents might be introduced together into a blight canker at one or more sites, and individual components might then multiply and migrate through the thallus at different rates, yielding regions containing different mixtures.

The variation among strains obtained from natural abnormal cankers in Italy might be related to the instability that many white European strains display in culture. Instability is expressed in three ways: variation among subcultures obtained from a given colony; sector formation within individual thalli; and segregation of colony types among single conidium isolates, colonies obtained from hyphal tips, and colonies obtained from mycelial fragments. Although information is limited at present, changes in dsRNA pattern sometimes accompany changes in morphological type (Anagnostakis and Day, 1979). Again, the basis for this variation is unknown, but could result from separation of components of mixtures, changes in relative populations of components of mixtures, or changes in populations of individual components when present alone in a thallus. Part of this instability also could be due to the formation and multiplication of mutant nuclei, the presence and segregation of genetically distinct nuclei within the fungal thallus, or to loss or incorporation of the genome of the cytoplasmic agent into the genome of the fungus (Van Alfen *et al.*, 1979).

2. The Abnormal Canker and Maintenance of Cytoplasmic Hypovirulent Strains

Persistent, superficial cankers are the most obvious reservoirs of CH agents. These cankers range from a few centimetres in length and width, to massive structures that encircle the trunks and extend more than a metre along their lengths (Fig. 5).

Abnormal cankers display ample evidence that the suscept responds vigorously to the pathogen. The tree confines the fungus to outer strata of bark by producing periderms beneath it. In some cankers, the thallus expands for extended periods, perhaps for many years. Exactly how the fungus accomplishes this is unknown. Discontinuities may occur in periderms at their extremities, where they obliquely cross strata of bark tissue. Such discontinuities may provide weak points in the host's line of defence that the fungus can breach. Besides erecting periderms, the plant responds with increased cambial activity and abnormal differentiation of inner bark tissues. Together these activities probably cause the slight swelling that characterizes abnormal cankers (Biraghi, 1953).

In time, the thallus stops growing and the process of expulsion goes gradually to completion. Canker size and form at containment probably depends on the original pathogenicity of the strain that initiates the infection, the size and complexity that the canker attains before CH agents are introduced into it, the types, debilitating capacities, and points the CH agents enter the canker and their rates of spread through the fungal thallus, and the ability of the tree to respond to the fungus. A strain need not be non-pathogenic to be contained and expelled by the host: it must be incapable of overcoming defences and breaching barriers erected against it by the suscept tissue that it occupies.

Maintenance in nature of highly debilitated hypovirulent strains, those with the least pathogenicity and most pronounced curative capacity, is more difficult to envision than maintenance of more pathogenic strains. Assuming that highly debilitated strains are not well adapted for a saprophytic existence, they have no obvious means for perpetuating themselves unaided. The most debilitated of these are non-pathogenic, and therefore unable to establish themselves and reproduce in the host. Unless vectors transmit highly debilitated strains from canker to canker, and the CH agents within them infect and multiply in the pre-existing thallus of the recipient canker, or these strains persist as saprophytes, they probably perish after infected bark containing them is shed. Thus, highly debilitated strains might be end-products of the biocontrol process, and be irrelevant to its operation.

The more pathogenic CH strains might maintain themselves by directly infecting healthy host tissue and producing abnormal cankers. Another possible mechanism for maintenance of CH agents is latency, either in the thallus at a level so low that they cause no detectable debilitation, or integrated into the genome of the fungus as suggested by Van Alfen *et al.* (1978). Stresses imposed on such strains during interactions with host tissue might allow the CH agents to multiply or become active, restoring the abnormal condition.

3. Natural Spread of Cytoplasmic Hypovirulent Strains and Their Accession to Dominance

Hypovirulent strains are at a competitive disadvantage with normal strains. Many European CH strains produce few conidia, and none are known to produce ascospores. The more debilitated a strain is, the less able it is to compete with normal strains.

There is little chance that hypovirulent strains will become dominant over normal strains if wind- and rain-borne conidia are the only significant form of inoculum available. Spread by vectors would seem to be essential if CH strains are to spread widely. The mycelial fans and colonized tissue, which seem to be the natural reservoirs of CH agents in Europe, may well serve as the source for spread by vectors with highly developed senses, such as arthropods (insects and arachnids), birds, or both. Cracks and crevices of abnormal cankers provide ideal shelter for insects, arachnids, and other small animal life, and the mycelial fans exposed there might serve as food for certain of them. Abnormal cankers in Italy contain abundant evidence that insects and arachnids visit and inhabit them. Cast exoskeletons, faeces, webs and other traces of their habitation have accumulated under the loose bark scales. Exposed stromata often appear to have been partly or totally devoured, probably by beetles or mites. Either these arthropods or insectivorous birds attracted to the cankers could serve as vectors. Both insects and birds acquire and transport inoculum of normal strains of *E. parasitica* (Craighead, 1912; Heald and Studhalter, 1914; Studhalter and Ruggles, 1915; Delen, 1975; Scharf and DePalma, 1982). Although birds probably are not important in the spread of CH strains in Italy, as hunting has kept their populations low, they may be important in the USA.

Vectors may serve three functions: (1) to initiate new abnormal cankers by inoculating uninfected tissue with pathogenic hypovirulent strains; (2) to introduce hypovirulent inoculum into normal cankers; and (3) to introduce additional CH agents into cankers that are already abnormal. Vectors may be essential to the perpetuation of highly debilitated forms as components of the biocontrol system, and they may be instrumental in creating these forms by introducing diverse agents into cankers.

In Italy, CH agents apparently are transmitted efficiently to normal cankers. In some localities where the incidence of normal blight was once high, normal cankers are now difficult to find. Stressed trees commonly emit chemicals that attract certain insects; perhaps vectors of hypovirulent strains are guided to incipient cankers by chemical attractants released by the fungus–suscept complex. Though the role of vectors in the natural biocontrol of *E. parasitica* has not been studied, such research might be most productive in Italy.

Even if adequate vectors and sources of CH inoculum were available, vegetative incompatibility among genotypes of the fungus could be an important constraint to spread of CH agents. Vegetative incompatibility was first described in *E. parasitica* by Grente and Sauret (1969a, b). Transmission of CH agents apparently requires cytoplasmic continuity between donor and recipient, and continuity must be maintained long enough for the agents to move from one thallus into the other. Anastomoses, or hyphal bridges, that interconnect cells of the same and different thalli, can provide this continuity. The genotypes of the individual strains largely determine whether viable anastomoses are established. Vegetatively compatible strains readily form anastomoses that persist for extended periods. Anastomoses formed between highly incompatible strains deteriorate (Grente and Sauret, 1969b; Anagnostakis, 1977), impeding migration of nuclei and cytoplasmic elements between thalli. Caten (1972) and Day (1978) suggested that the biological function of vegetative incompatibility in fungi might be to limit spread of cytoplasmically transmitted infectious agents in a population. Alternatively, Hartl *et al.*, (1975) suggested that it prevents exploitation of heterokaryons by non-adaptive nuclei.

Although vegetative incompatibility appears to hinder transmission of CH agents, it does not absolutely prevent it. Often transmission between incompatible strains can be achieved via multiple confrontations in suscept tissue (Grente and Berthelay-Sauret, 1979). In nature, vectors might foster multiple confrontations.

The basis for the imperfections in vegetative incompatibility barriers is not understood. CH agents may weaken incompatibility barriers by altering the cytoplasm of the donor strain so that anastomoses remain alive long enough for the agents to pass. Alternatively strains carrying a greater diversity or a larger population of CH agents may be better able to transmit at least a few copies of the agent(s) before anastomoses are blocked. A strain made universally compatible by CH agents might be ideal.

Whether vegetative incompatiblity groups were numerous among strains of *E. parasitica* in Italy before CH became established is not known. Woodruff's account of chestnut blight in Italy during the early 1940s (Woodruff, 1946) suggests that perithecia and ascospores were common then in at least some areas. If several genotypes of the fungus were present at the outset of the epidemic, and if sexual reproduction were occurring at a high frequency, recombinant genotypes probably would include many vegetative compatibility groups. Unfortunately, we will never know what the vegetative incompatibility situation was during that time.

The present status of vegetative incompatibility in the Italian population of *E. parasitica* has not been well characterized. Fragmentary information obtained from samples taken by the author in August, 1978, from trees in the Piedmont (northern), Tuscany (north central), and Campania (southern) regions suggest that there are now relatively few compatibility types present. At the time these samples were taken in these areas of Italy, normal cankers were infrequent and normal cankers containing perithecia were rare, suggesting that sexual reproduction is now rare.

The genetics of vegetative compatibility in Italy may have always been simple, or perhaps have become simpler since CH agents became established in the population. It seems improbable that a pronounced simplification of a complex vegetative incompatibility system could have occurred over a 30- to 40-year period, but perhaps this has happened in Italy. European CH agents greatly reduce or eliminate sexual reproduction in *E. parasitica*. Gradual loss of the sexual stage as CH agents spread would gradually decrease production of new compatibility types. The number of types could then decline. In Italy, where much of the chestnut is coppiced on a 7-to 20-year rotation, many compatibility groups could have been eliminated by man. If vectors spread CH agents efficiently, the probability that protoperithecia would be spermatized by conidia from other compatibility groups would be low. Thus the restoration of a complex compatibility system in the population would be difficult.

G. Possible Origin of Hypovirulence in *Endothia parasitica*

The origins of CH agents that affect *E. parasitica* is unknown, and available evidence can be used to argue either for ancient or recent acquisition. A thorough study of the fungus population in the Orient, its apparent centre of evolution, could provide an answer. If CH agents were absent from oriental populations, or if agents present in them were distinct from European and North American forms, acquisition after the fungus was introduced into the western hemisphere could be considered likely.

Hypovirulence in *Endothia* may have both nuclear and cytoplasmic components. The following hypothesis might account for its development.

Current geophysical theory based on plate tectonics (Dott and Batten, 1976) indicates that two continents, Laurasia in the northern hemisphere and Gondwanaland in the southern hemisphere, merged into one supercontinent, Pangea, by the process of continental drift. During the time Pangea persisted, the angiosperms originated, dispersed over the land mass, and began divergent, adaptive evolution. Physical or climatological barriers developed that limited movement of species in the north-south direction but not in the east–west direction, thus, after an initial period of unrestricted migration, dispersal of species was possible only laterally in Laurasian and Gondwanalandic provinces. This led to the evolution of distinct angiosperm flora in the northern and southern hemispheres. The Fagales, a large order of angiosperms that includes oaks, beeches, chestnuts, birches, alders, hornbeams and others, evolved in Laurasia after the barriers were established. This interpretation accounts for the confinement of nearly all members of the Fagales to the northern hemisphere and the close general similarities among North American, European and Asian flora (Graham, 1972). Subsequent migrations of members of this order into the southern hemisphere occurred by complex routes and processes (Schuster, 1976). It is likely that the ancestor of *Endothia* established its parasitic associates early in their evolution.

With time, portions of the Laurasian flora became separated and continued to evolve in isolation. The periodic advances and retreats of glaciers that accompanied the passage of the ice ages influenced the evolution of the isolated components differently, as did changes in climate as continents drifted and interacted on their paths across the surface of the earth. These changes surely contributed to the extinction of many forms of life that were unable to migrate southward in advance of the glaciers or adapt adequately to changing climate. This left a residue of related, similar but distinct, species on the continents as they presently exist. Within the separated biota, the associations between suscepts and pathogens probably evolved over the millions of years since their inception such that balanced pathosystems were maintained.

The original *Castanea–Endothia* pathosystem apparently was separated into at least three subsystems during its evolution: one in eastern North America, one in western Europe and Asia Minor, and one in Asia. The subsystems in North America and Europe apparently evolved similarly such that the weak resistance of the respective *Castanea* components were in balance with the weak pathogenicities of the respective *Endothia* components. The components of the Asian subsystem also evolved more or less balanced relationships, but with higher levels of resistance and pathogenicity on the parts of the suscepts and pathogen, respectively. The higher level of pathogenicity in the Asian fungus appears to be associated with a capacity to produce mycelial fans in suscept tissue, a characteristic found so far, among *Endothia* species, only in *E. parasitica* (Shear *et al.*, 1917). As each of the isolated subsystems evolved, the fungus components may have independently acquired and adapted to distinct parasites of their own, agents that we now recognize in *E. parasitica* as CH agents.

North American, European and Asian *Castanea* species did not evolve effective mechanisms of reproductive isolation and hybridize freely when allowed to interact (Jaynes, 1972). Thus, they are subject to rapid convergent evolution if brought into close proximity. Certain of their associated *Endothia* species may also have hybridized. In the time since *E. parasitica* entered North America and Europe, it may have interacted with related species native to these regions and exchanged nuclear and cytoplasmic genetic elements. These native *Endothia* species may have been the source of CH agents present in North American and European strains of the fungus. Thus, two independent processes occurring simultaneously might be attenuating the virulence of this Asian fungus: (1) the development of CH by infection with internal parasites from native North American and European relatives, and (2) the development of nuclear hypovirulence by hybridization, heterokaryon formation, or both.

Since levels of virulence often appear to be controlled by many genes (Burnett, 1975), hybridization of forms with disparate levels of virulence should yield recombinant genotypes with a wide range of virulence. Most of these genotypes should confer levels of virulence below that of the most virulent parent. This occurred when aggressive and non-aggressive strains of *Ceratocystis ulmi* were hybridized (Brasier and Gibbs, 1976; Brasier, 1977). If such hybridization occurred frequently in the pathogen population, the average level of virulence should gradually decline. Heterokaryon formation between individuals with different levels of virulence might have similar effects. Simple examples of these processes, with and without involvement of CH agents, can be depicted symbolically for the interactions between *E. parasitica* and the related European *Endothia* (Figs. 9 and 10).

A third type of interaction not depicted in Figs. 9 and 10 is transmission of CH agents from the European to the Asian fungus by

(a) $E_0 + E_E \rightleftharpoons E_{0+E} \rightarrow E_{0E} \rightarrow E_{0E/2}$

(b) $E_0 + E_E \rightarrow E_{0E} \rightarrow E_{0E/2}$

Fig. 9. Equations representing (a) heterokaryon formation between *Endothia parasitica* and the native European *Endothia* species, E_0 and E_E, respectively, followed by diploidization and meiosis and (b) fertilization followed by meiosis, both interactions occurring in the absence of CH agents.

$$\text{(a)}\quad E_0 + E_E(H_{E_1\ldots E_n}) \rightleftharpoons E_{0+E}(H_{E_1\ldots E_n}) \rightarrow E_{0E}(H_{E_1\ldots E_n}) \rightarrow E_{0E/2}(H_{E_1\ldots E_n})$$

$$E_E(H_{E_1\ldots E_n}) \rightleftharpoons E_E; \quad E_{0+E}(H_{E_1\ldots E_n}) \rightleftharpoons E_0(H_{E_1\ldots E_n}) + E_E; \quad E_{0E/2}(H_{E_1\ldots E_n}) \rightleftharpoons E_{0E/2}$$

$$\text{(b)}\quad E_0 + E_E(H_{E_1\ldots E_n}) \rightarrow E_{0E}(H_{E_1\ldots E_n}) \rightarrow E_{0E/2}(H_{E_1\ldots E_n})$$

$$E_E(H_{E_1\ldots E_n}) \rightleftharpoons E_E; \quad E_{0E/2}(H_{E_1\ldots E_n}) \rightleftharpoons E_{0E/2}$$

Fig. 10. The same two processes depicted in Fig. 9 with CH agents present in the European *Endothia* component. These equations depict faithful transmission of one or more CH agents. If transmission of a mixture of CH agents were not faithful, i.e. if they segregated, the situation would be far more complex. Symbols: E_0 = genome of *E. parasitica*; E_E = genome of the native European *Endothia* species; E_{0+E} = heterokaryon formed between the two; E_{0E} = transitory diploid; $E_{0E/2}$ = hybrid genome; $H_{E_1\ldots E_n}$ = one or more CH agents from the European species.

anastomosis without heterokaryon formation or hybridization. Once the agents were introduced into the alien fungus and established in its population, no further interaction with the native species would be necessary. However, if the two species interacted frequently, the agents would become more rapidly and completely distributed within the pathogen population. Similar interactions between *E. parasitica* and the related North American species can be envisioned.

Bonifacio and Turchetti (1973) suggested that heterokaryon formation between highly pathogenic and non-pathogenic forms might account for the behaviour of hypovirulent strains they obtained from abnormal cankers in north central Italy. They observed that single conidium isolates of certain white strains (Fig. 6b) segregated into four classes: normal strains, i.e. nonsegregating; highly pathogenic forms typical of *E. parasitica* as described by Shear *et al.*, (1917) (Fig. 6a); non-segregating pigmented hypovirulent isolates (Fig. 6d), forms that Grente and Sauret (1969b) described as "JR" types; segregating white isolates with moderately stable cultural characteristics (Fig. 6b); and segregating intermediate isolates that frequently produced sectors when cultured on laboratory media (Fig. 6c). Both normal and pigmented hypovirulent strains were stable; they yielded only single conidium isolates like the respective parent. Anastomoses between normal isolates and pigmented hypovirulent isolates yield white and intermediate types (Bonifacio and Turchetti, 1973; Grente and Berthelay-Sauret, 1979; Elliston, unpublished). Bonifacio and Turchetti (1973) isolated both of the stable types from abnormal cankers. The behaviour of these is consistent with a hypothesis based on heterokaryon formation. A phenomenon that is at least superficially similar occurs in *Fusarium oxysporun* f. sp. *lycopersici*, in which formation of heterokaryons between nuclei having distinct genomes has been documented (Sanches *et al.*, 1975, 1976). Perhaps the stable, pigmented, non-pathogenic form in Italy is the native European *Endothia* species.

Other observations are difficult to reconcile with the heterokaryon hypothesis. White strains obtained from interactions between normal and pigmented hypovirulent strains do not regularly yield conidia that produce colonies of the pigmented type. Because conidia of *E. parasitica* typically are mononucleate (Anderson, 1913; Bonifacio and Turchetti, 1973), if stable heterokaryons were formed, the pigmented type should not be difficult to recover by this method. That they are often difficult to obtain suggests either that nuclear genes do not control the pigmented phenotype, or nuclei contributed by the pigmented strain do not persist or, they become a very small proportion of the population of nuclei. The peculiar characteristics of the pigmented type from Europe will not be understood until isolates free of dsRNA are obtained. So far, all of the representatives of this type that have been tested contain dsRNA (Day *et al.*, 1977; Dodds, 1980a; Elliston, unpublished).

The key components of the main hypothesis for the origin of hypovirulence in *E. parasitica* have not been accounted for, i.e. the counterparts of *E. parasitica* that evolved in North America with *C. dentata* and in Europe with *C. sativa*. Of the species of *Endothia* known to occur in these regions, *Endothia radicalis* (Schw.) De Not. (*E. fluens* (Sow.) S. and S.) is the most likely candidate both in North America and in Europe. The evidence for this is as follows.

Endothia radicalis (Schw.) De Not. occurred as a weak parasite or as a saprophyte on *C. sativa* and some of its allies throughout southern Europe. In North America, this fungus was found principally on *C. dentata* within the southern two-thirds of the natural range of the tree. The natural range of *E. radicalis* in North America conformed strictly with the range of *C. dentata*, indicating a close association between these species (Shear *et al.*, 1917).

During the early phases of the chestnut blight epidemic in North America, a controversy developed over the identity of this fungus and its relationship to the fungus that was causing the blight. Farlow (1912) and Clinton (1912b) first recognized the close similarities between the blight fungus, the *E. radicalis* found on chestnut in Italy, and a species found by Schweinitz in the southern Appalachians early in the nineteenth century. Saccardo and Von Hohnel, European botanists, compared the chestnut blight fungus from North America with the European *E. radicalis* and found them to be indistinguishable on morphological grounds (Farlow, 1912). Clinton recognized minor morphological and cultural differences between the blight fungus and the North American and European species, but he considered them significant only at the variety level. He named the blight fungus *Endothia gyrosa* var. *parasitica* (Murr.) Clint. and considered the other two to be *E. gyrosa* (Schw.) Fr., a synonym for *E. radicalis* (Schw.) De Not. (Clinton, 1912a, b, 1913). Shear (1912) examined *E. radicalis* (Schw.) De Not. in many areas of Europe and concluded initially that it was virtually identical with the chestnut blight fungus. He suggested that Europe might have been the source of the pathogen. He later retracted this suggestion (1913) and, after further study of many samples of *E. radicalis* from Europe and North America, concluded that it could be separated from the blight fungus on the basis of the differences that Clinton did not think were sufficient to separate the organisms into different species (Shear and Stevens, 1913b). Later Shear *et al.*, (1917) assigned the name *E. fluens* (Sow.) S. and S. to the species native to Europe and North America. Anderson and Anderson, (1912a, b), also studied *E. radicalis*, known to them as the Connellsville fungus since they first found it near Connellsville, Pennsylvania, and named it *E. virginiana* Anders., since they were unaware that it had been described previously. In the same paper, they suggested the name *E. parasitica* (Murr.) P. and H. And. for the blight fungus. Finally, Orsenigo (1951a, b) compared a large collection of *Endothia* species in detail and concluded that the differences between specimens of *E. radicalis* (Schw.) De Not. and *E. parasitica* were so minor that the blight fungus could not be assigned to a separate species. He referred both to *E. radicalis* (Schw.) De Not. as subspecies.

Evidence presently available that supports the hypothesis that *E. parasitica* and *E. radicalis* may have interacted significantly in North America and Europe is circumstantial. In North America, *E. radicalis* apparently has disappeared (J. Stipes, personal communication). At the time that the blight fungus was being studied intensively and the controversy was raging over the identity of *E. radicalis*, this native fungus occurred abundantly from southern Pennsylvania southward within the range of *C. dentata*. It was found chiefly on exposed roots of chestnut and several oak species, on recently cut stumps of chestnut, and in injured bark tissue (Anderson and Anderson, 1912b; Shear *et al.*, 1917). Its apparent disappearance since then could be related to the decimation of its principal host, *C. dentata*. However, remnants of the original *C. dentata* population persist as sprouts scattered throughout its natural range, often in poor condition due to repeated infection and suppression. In this condition, it should be an ideal host for *E. radicalis*, and one might expect this fungus to be thriving, if not on chestnut, on some of its other hosts that replaced chestnut.

Clinton's observation (Clinton, 1924) that the epidemic slowed as the fungus entered the southern Appalachians is also unexpected. Stevens (1917) compared the rates of growth and development of *E. parasitica* in chestnut at many points between the southern and northern limits of its range. Based on the results of these experiments, he predicted that the fungus would move swiftly through the southern chestnut forests, because it would encounter conditions most favourable for its development there.

These observations are consistent with the hypothesis that *E. parasitica* and *E. radicalis* may have interacted in significant ways in North America after they came into contact. *Endothia radicalis* may have been absorbed by *E. parasitica* by hybridization and, as one consequence, the virulence of *E. parasitica* may have declined somewhat but not enough to allow its weakly resistant suscept, *C. dentata*, to withstand its attack.

If hybridization between these species occurred, one might predict that the host range of the fungus would have been broadened to include oaks or other members of the Fagales, because hybridization between species of pathogenic fungi often expand or alter their host ranges as well as reduce their levels of virulence (Burnett, 1975). Anderson and Rankin (1914) and Clinton (1913) noted that *E. parasitica* occurred on freshly cut oaks during the early stages of the epidemic. Biraghi (1950c, 1951) reported it on oaks in Italy. Orsenigo (1951b) suggested that a broadening of the host range of *E. parasitica* in Europe may have occurred as a consequence of mutation or hybridization. This might also have happened in North America, but it would be difficult to determine. The butt swell of scarlet oak (*Quercus coccinia*) in the southern Appalachians has been attributed to infection by *E. parasitica* on the basis of cultural characteristics and production of mycelial fans in infected tissue (Ham, 1967). Likewise, canker diseases of post oak *Q. stellata* (Clapper *et al.*, 1946) and live oak, *Q. virginiana*, (May and Davidson, 1960; Batson and Witcher, 1968) have been attributed to this fungus. Although *E. radicalis* did not produce mycelial fans in host tissue (Anderson and Anderson, 1912a; Orsenigo, 1951b), its hybrids with *E.*

parasitica might well produce them. Reports that greater numbers of cankers were required to kill a sprout after the fungus had been present in an area for many years (Aughanbaugh, 1930) suggest that the virulence of the fungus might have declined.

The discovery of North American CH strains of *E. parasitica* could also be a consequence of interaction between *E. parasitica* and *E. radicalis*. The absence of widespread CH in the North American population of *E. parasitica* could have several causes. Some American CH agents do not prevent the fungus from producing normal ascospores (Elliston, unpublished). Thus, the fungus has a mechanism for cleansing itself of these CH agents. Another important factor is the high frequency of sexual reproduction in the North American *E. parasitica* population, which provides many recombinant genotypes and a greater potential for maintaining numerous barriers to transmission of cytoplasmic agents. Also, if CH agents occurred only sporadically in the *E. radicalis* population, they may have had little impact on the pathogenicity of the hybrids.

Similar circumstantial evidence can be cited for interaction between *E. parasitica* and *E. radicalis* in Europe. Biraghi (1953b) observed that abnormal cankers were found almost simultaneously in many areas of Italy where blight was present longest. Palenzona (1978) expressed the view that the blight fungus gradually loses its virulence after it enters a new area and before the abnormal white forms appear.

In Europe, *Endothia radicalis* occurred as a weak pathogen or saprophyte on exposed roots, branch stubs, injured bark, and weakened trees of chestnut and several other hosts (Shear, 1912b; Shear *et al.*, 1917). Heavy blight infection and cutting of diseased trees should have provided ideal conditions for *E. radicalis* to increase and surely provided ample opportunities for interactions between it and *E. parasitica*. Because *E. radicalis* was widely distributed in Italy and southern Europe, this interaction could have had much greater impact than in North America, where the blight epidemic built up momentum in an area free of *E. radicalis*. The cytoplasmic agents in the European *E. radicalis* population also might have been distinct from those in North America and better able to reduce sexual reproduction.

Though the hypothesis that *E. parasitica* interacted with *E. radicalis* in North America and Europe, and that hypovirulence developed in the *E. parasitica* population as a result is consistent with the observations cited above and is perhaps the simplest explanation, experimental matings between North American, European, and oriental strains of *E. parasitica* and *E. radicalis* have not yet been made. If, when tested in sufficient numbers, *E. parasitica* and the North American and European *E. radicalis* fail to hybridize, or if CH agents cannot be transmitted between them, the hypothesis will have to be abandoned.

An alternative hypothesis is that *E. parasitica* entered North America and Europe carrying distinct CH agents, and that in Europe the CH agents became well established in the population, whereas in North America they did not. If this hypothesis were correct, the respective CH agents should occur in the Asian population of the fungus. A thorough study of that population would be the only way to determine the merits of this alternative hypothesis.

Another possibility is that *E. parasitica* entered North America and Europe free from CH agents and acquired them by chance interactions with unrelated or distantly related organisms it encountered. Again, it should be possible to find the organisms in North America and Europe that contain these same agents and effect the necessary experimental interactions. Organisms associated with bark of American and European chestnut would be the most likely candidates for such studies.

IV. HYPOVIRULENCE IN OTHER PATHOSYSTEMS

Another possible example of hypovirulence involves *Ulmus* spp. and *Ceratocystis ulmi*, the Dutch elm disease pathosystem. Until 1971, only one type of strain of *C. ulmi* was recognized. In 1971, in England, a new, more pathogenic type was discovered causing serious new outbreaks of Dutch elm disease (Gibbs and Brasier, 1973). Some species of elm and hybrid elms that previously had been little affected by the disease were killed rapidly by the new strain. This strain grew rapidly and produced fluffy colonies on 2% Oxoid malt extract agar at 18°C. The endemic strain grew significantly less rapidly and produced colonies with mycelium appressed to the agar surface, giving the colonies a waxy appearance. The new, more pathogenic type was designated aggressive, and the endemic type was designated non-aggressive. Intergrading types were not found.

After its initial discovery in England, the aggressive strain was found causing serious outbreaks of Dutch elm disease in France, Germany, the Netherlands, Italy, European Russia, Iran and North America (Gibbs, 1978). Circumstantial evidence suggested that this strain entered England on elm logs imported from North America (Brasier and Gibbs, 1973).

Atanasoff (1973) suggested that the non-aggressive forms might be infected with hyperparasites, mycoviruses in particular. Hollings *et al.* (1974, 1975) were unable to detect virus-like particles or dsRNA in strains of either type. Progeny of crosses between representatives of the aggressive and non-aggressive types displayed cultural characteristics and pathogenicity in differential suscepts that spanned those of the two parents (Brasier and Gibbs, 1976; Brasier, 1977). Differences between the two types were polygenic rather than monogenic, discounting the hypothesis that the aggressive form might have arisen by simple mutation.

Studies of the distribution of aggressive and non-aggressive types in North America support the view that the aggressive type has been present longest in the midwestern United States (Gibbs *et al.*, 1979). Whether it arose there or was introduced from the centre of evolution of *C. ulmi*, probably Asia, is unknown.

Although Gibbs and Brasier's results suggest that cytoplasmic agents probably do not play a role in determining the lower pathogenicity of the non-aggressive type of *C. ulmi*, Pusey and Wilson (1979) found dsRNA in isolates of this fungus. They examined isolates with known pathogenicities, using polyacrylamide gel electrophoresis and serological tests, and found dsRNA in all non-aggressive strains, strains intermediate in aggressiveness, and in a few of the aggressive strains. The patterns of dsRNA bands for non-aggressive strains were distinct from those for aggressive strains and strains intermediate in aggressiveness. What role these agents might play in determining the degree of aggressiveness of the strains has not been determined.

If cytoplasmic agents are responsible for loss of aggressiveness in *C. ulmi*, they provide a potential means for reducing the virulence of the aggressive types and restoring greater balance in parts of the pathosystem that involve some of the less susceptible European elm species and hybrids. They would have no obvious use in controlling the disease in *U. americana* because aggressive and non-aggressive strains do not show differential pathogenicity in this species. The difficulty of using cytoplasmic agents in the Dutch elm disease pathosystem is greater than in the *Castanea–Endothia* pathosystem because the elm pathogen colonizes the suscept rapidly, is cryptic, and therefore difficult to locate in the infected tree. This pathosystem does have the advantage of a well-developed vector relationship that might be used to advantage. Prospects for natural hybridization between the aggressive and non-aggressive forms to generate nuclear hypovirulence also are dim because effective barriers to genetic interaction exist between the two forms (Brasier, 1977).

An example of CH in a basidiomycete, *Rhizoctonia solani*, was described by Castanho *et al.* (1978a, b, c). An isolate of this fungus that had pronounced cultural abnormalities and subnormal virulence contained dsRNA. The dsRNA was transmitted to normal subcultures by hyphal anastomosis, and the recipient then became diseased. The diseased isolate reduced the incidence of damping-off in sugar-beet seedlings, when incorporated into the soil with the seed. Unfortunately, the effect was short-lived, and applied only to *R. solani* with the same genotype as the infected strain. In *R. solani*, diseased strains apparently are rather common; Castanho and Butler (1978a) reported thirty of them.

CH agents may be more common in fungi than is apparent, particularly if, to be detected, they must be introduced into organisms that have not evolved adaptations to them. Strains of fungi containing CH agents might go undetected or be discarded as contaminants or unidentifiable nuisances if the agents mask the identity of the fungus.

V. CONCLUDING REMARKS

Over the course of the past several thousand years, and particularly during the last 500 years, man has served as an extraordinary vehicle for moving genetically related and unrelated components of the earth's flora from one geographically isolated area to another. Most of this activity has had beneficial effects, from man's point of view, and has been relatively harmless to native flora. But, occasionally, it has seriously upset balanced pathosystems. Hypovirulence phenomena may provide natural solutions to such imbalances.

REFERENCES

Anagnostakis, S. L. (1977). *Expl. Mycol.* **1,** 306–316.

Anagnostakis, S. L. and Day, P. R. (1979). *Phytopathology* **69,** 1226–1229.

Anderson, P. J. (1913). *Pennsylvania Chestnut Tree Blight Commission Bull. 7.*

Anderson, P. J. and Anderson, H. W. (1912a). *Phytopathology* **2,** 204–210.

Anderson, P. J. and Anderson, H. W. (1912b). *Phytopathology* **2,** 261–262.

Anderson, P. J. and Rankin, W. H. (1914). *Cornell Agr. Exp. Sta. Bull.* **347,** 530–620.

Anonymous (1915). *The Publications of the Pennsylvania Chestnut Tree Blight Commission*, Harrisburg.

Atanasoff, D. (1973). *Phytopath. Z.* **78,** 182–186.

Aughanbaugh, J. E. (1930). *Penn. Dept. Forests Wat. Res. Circ.* **1,** 1–18.

Bagnaresi, U. (1956). *Centro di Studio sul Castagno*, **3,** 8–41.

Batson, W. E. and Witcher, W. (1968). *Phytopathology* **58,** 1473–1475.

Berthelay-Sauret, S. (1973). *Ann. Phytopath.* **5,** 318.

Biraghi, A. (1946). *L'Italia Agric.* **7,** 406–412.
Biraghi, A. (1950a). *L'Italia for. Mont.* **5,** 18–21.
Biraghi, A. (1950b). *Boll. Staz. Patol. veg. Roma* VII, III, **5,** 167–171.
Biraghi, A. (1950c). *Ann. Sper. Agr.* **4,** 109–118.
Biraghi, A. (1951). *L'Italia for. mont.* **6,** 15–16.
Biraghi, A. (1953a). *IUFRO, XI Cong. Proc.*, 643–645.
Biraghi, A. (1953b). *Boll. Staz. Patol. veg. Roma,* **XI,** 149–157.
Bonifacio, A. and Turchetti, T. (1973). *Ann. Accad. Ital. Sci. For.* **22,** 111–131.
Brasier, C. M. (1977). *Trans. Br. Mycol. Soc.* **68,** 45–52.
Brasier, C. M. and Gibbs, J. N. (1973). *Nature, Lond.* **242,** 607–609.
Brasier, C. M. and Gibbs, J. N. (1976). *Ann. Appl. Biol.* **83,** 31–37.
Burnett, J. H. (1975). "Mycogenetics", J. Wiley and Sons, London.
Castanho, B., Butler, E. E. and Shepherd, R. J. (1978a). *Phytopathology* **68,** 1515–1519.
Castanho, B. and Butler, E. E. (1978b). *Phytopathology* **68,** 1511–1514.
Castanho, B. and Butler, E. E. (1978c). *Phytopathology* **68,** 1505–1510.
Caten, C. E. (1972). *J. Gen. Microbiol.* **72,** 221–229.
Clapper, R. B., Gravatt, J. P. and Stout, D. C. (1946). *Pl. Dis. Reptr.* **30,** 381.
Clinton, G. P. (1912a). *Phytopathology* **2,** 265–269.
Clinton, G. P. (1912b). *Science, N.Y.* **36,** 907–914.
Clinton, G. P. (1913). *Conn. Agr. Exp. Sta. Ann. Rept.* 1912, 359–453.
Clinton, G. P. (1924). *New Engl. Fms.* **54,** 1–9.
Craighead, F. C. (1912). *Science, N.Y.* **36,** 825.
Day, P. R. (1978). *In* "Genetics and Morphogenesis in the Basidiomycetes", (M. N. Schwalb and P. G. Miles, eds), pp. 67–79. Academic Press, New York and London.
Day, P. R. and Dodds, J. A. (1979). *In* "Viruses and Plasmids of Fungi", (P. A. Lemke, ed.), pp. 201–237. Marcel Dekker, New York.
Day, P. R., Dodds, J. A., Elliston, J. E., Jaynes, R. A. and Anagnostakis, S. L. (1977). *Phytopathology* **67,** 1393–1396.
Delen, N. (1975). *J. Turk. Phytopath.* **4,** 93–113.
Dodds, J. A. (1979). *In* "Proceedings of the American Chestnut Symposium", (W. L. MacDonald, F. C. Cech, J. Luchok and C. Smith, eds), pp. 108–109. West Virginia University Press, Morgantown.
Dodds, J. A. (1980a). *Phytopathology* **70,** 1217–1220.
Dodds, J. A. (1980b). *Virology* **107,** 1–12.
Dodds, J. A. and Elliston, J. E. (1978). *Third Int. Cong. Plant Pathol.*, Abstr. 72.
Dott, R. and Batten, R. (1976). "Evolution of the Earth." McGraw-Hill, New York.
Elliston, J. E. (1977). *Proc. Am. Phytopath. Soc.* **4,** 111.
Elliston, J. E. (1979). *In* "Proceedings of the American Chestnut Symposium", (W. L. MacDonald, F. C. Cech, J. Luchok and C. Smith, eds), pp. 95–100. West Virginia University Press, Morgantown.
Elliston, J. E., Jaynes, R. A., Day, P. R. and Anagnostakis, S. L. (1977). *Proc. Am. Phytopath. Soc.* **4,** 83.

Elliston, J. E., McCarroll, D. and Wooding, B. (1979). *IX Int. Cong. Plant Protection*, Abstr. 633.
Eynard, I., Honda, N. and Koma, S. (1966). *Atti Convegno Int. sul Castagno*, 90–106.
Fairchild, D. (1913). *Science N. Y.* **38,** 297–299.
Farlow, W. G. (1912). *Science N.Y.* **35,** 717–722.
Gibbs, J. N. (1978). *A. Rev. Phytopathol.* **16,** 287–307.
Gibbs, J. N. and Brasier, C. M. (1973). *Nature, Lond.* **241,** 381–383.
Gibbs, J. N., Houston, D. R. and Smalley, E. G. (1979). *Phytopathology* **69,** 1215–1219.
Graham, A. (1972). "Floristics and Paleofloristics of Asia and Eastern North America", Elsevier, Amsterdam and New York.
Grente, J. (1965). *C. r. hebd. Seanc. Acad. Agric. Fr.* **51,** 1033–1037.
Grente, J. (1975). *Ann. Phytopath.* **7,** 216–218.
Grente, J. and Sauret, S. (1969a). *C. r. hebd. Seanc. Acad. Sci. Paris, D.* **268,** 2347–2350.
Grente, J. and Sauret, S. (1969b). *C. r. hebd. Seanc. Acad. Sci. Paris, D.* **268,** 3173–3176.
Grente, J. and Berthelay-Sauret, S. (1979). *In* "Proceedings of the American Chestnut Symposium", (W. L. MacDonald, F. C. Cech, J. Luchok and C. Smith, eds), pp. 30–34. West Virginia University Press, Morgantown.
Ham, D. L. (1967). "The Etiology of *Endothia parasitica* on scarlet oak." MS Thesis, Duke Univ.
Hartl, D. L., Dempster, E. R. and Brown, S. W. (1975). *Genetics, Princeton.* **81,** 553–569.
Heald, F. D. and Studhalter, R. A. (1914). *J. agric. Res.* **2,** 405–422.
Hollings, M., Stone, O. M., Barton, r. j. and Atkey, P. T. (1974). *Rep. Glasshouse Crops Res. Inst. for 1973*, p. 124.
Hollings, M., Stone, O. M., Barton, R. J. and Atkey, P. T. (1974). *Rep. Glasshouse Crops Res. Inst. for 1973*, p. 124.
Hu, H. H. and Chaney, R. W. (1940). "A Miocine Flora from Shantung Province, China", *Carnegie Inst. of Washington Publ. No. 507.*
Jaynes, R. A. (1972). *U.S.D.A. For. Serv. Res. Paper WO-17.*
Jaynes, R. A., Anagnostakis, S. L. and Van Alfen, N. K. (1976). *In* "Perspectives in Forest Entomology", (J. E. Anderson and H. Kaya, eds), pp. 61–70. Academic Press, New York and London.
Kajiura, M. (1955). "Chestnut of Japan." Mimeo, 20 pp.
Kuhlman, E. G. (1980). *In* "Plant Disease—An Advanced Treatise", (J. G. Horsfall and E. B. Cowling, eds) Vol V, pp. 363–380. Academic Press, New York and London.
Lemke, P. A. and Nash, C. H. (1974). *Bact. Rev.* **38,** 29–56.
Lesemann, D. E. and Koenig, R. (1977). *Phytopath. Z.* **89,** 161–169.
May, C. and Davidson, R. W. (1960). *Pl. Dis. Reptr.* **44,** 754.
Mittempergher, L. (1979). *In* "Proceedings of the American Chestnut Symposium", (W. L. MacDonald, F. C. Cech, J. Luchok and C. Smith, eds), pp. 34–37. West Virginia University Press, Morgantown.

Moffitt, E. M. and Lister, R. M. (1975). *Phytopathology* **65,** 851–859.

Mullick, D. B. (1977). *Recent Adv. Phytochem.* **11,** 395–439.

Murrill, W. A. (1906a). *Torryea* **6,** 186–189.

Murrill, W. A. (1906b). *J. N.Y. bot. Gdn.* **7,** 143–153.

Murrill, W. A. (1908). *J. N.Y. bot. Gdn.* **9,** 23–30.

Nelson, R. R. (1979). *In* "Plant Disease—An Advanced Treatise", (J. G. Horsfall and E. B. Cowling, eds), Vol. IV, pp. 23–46. Academic Press, New York and London.

Orsenigo, M. (1951a). *Phytopath. Z.* **18,** 196–209.

Orsenigo, M. (1951b). *Phytopath. Z.* **18,** 210–220.

Palenzona, M. (1978). "Study and Action Program for the Use, Regeneration and Transformation of Chestnut Forests in Piedmont", *Tech. Rept. Giacomo Piccarolo Natl. Inst. for Woody Plant, Torino, Italy.*

Pavari, A. (1949). *Unasylva* **3,** 8–13.

Pusey, P. L. and Wilson, C. L. (1979). *IX Int. Cong. Plant Protection.* Abstr. 333.

Sanchez, L. E., Leary, J. V. and Endo, R. M. (1975). *J. Gen. Microbiol.* **87,** 326–332.

Sanchez, L. E., Leary, J. V. and Endo, R. M. (1976). *J. Gen. Microbiol.* **93,** 219–226.

Scharf, C. S. and DePalma, N. K. (1982). *Can. J. Zool.* (in press).

Schuster, R. M. (1976). *In* "Origin and Early Evolution of Angiosperms", (C. B. Beck, ed.), pp. 48–138. Columbia Univ. Press, New York.

Shear, C. L. (1912). *Phytopathology* **2,** 211–212.

Shear, C. L. (1913). *Phytopathology* **3,** 61.

Shear, C. L. and Stevens, N. E. (1913a). *Science N. Y.* **38,** 295–297.

Shear, C. L. and Stevens, N. E. (1913b). *Bur. Plant Ind. Circ.* 131.

Shear, C. L. and Stevens, N. E. (1916). *Science N. Y.* **43,** 173–176.

Shear, C. L., Stevens, N. E. and Tiller, R. J. (1917). *USDA Bull.* 380.

Smith, E. H. and Pimentel, D. (1978). "Pest Control Strategies." Academic Press, New York and London.

Steen, E. B. (1971). "Dictionary of Biology", Barnes and Noble, New York.

Stevens, N. E. (1917). *Am. J. Bot.* **4,** 1–32.

Studhalter, R. A. and Ruggles, A. G. (1915). *Penn. Dept. For. Bull.* **12.**

Turchetti, T. A. (1979). *In* "Proceedings of the American Chestnut Symposium", (W. L. MacDonald, F. C. Cech, J. Luchok and C. Smith, eds), pp. 92–94. West Virginia University Press, Morgantown.

Van Alfen, N. K., Jaynes, R. A. Anagnostakis, S. L. and Day, P. R. (1975). *Science N. Y.* **189,** 890–891.

Van Alfen, N. K., Jaynes, R. A. and Bowman, J. T. (1978). *Phytopathology* **68,** 1075–1079.

Van Alfen, N. K., Bowman, J. T. and Simmons, J. R. (1979). *In* "Proceedings of the American Chestnut Symposium", (W. L. MacDonald, F. C. Cech, J. Luchok and C. Smith, eds), pp. 106–108. West Virginia University Press, Morgantown.

Woodruff, J. B. (1946). *J. Am. Arboricult.* **6,** 8–16.

Woods, A. (1974). "Pest Control—A Survey", J. Wiley and Sons, New York.

2

SPECIFIC RECOGNITION IN GENE-FOR-GENE HOST–PARASITE SYSTEMS

N. T. Keen

Department of Plant Pathology, University of California, Riverside, California

I. Introduction
II. Requirements of Successful Pathogens
III. The Hypersensitive Reaction (HR)
  A. The Hypersensitive Reaction and Phytoalexins
  B. Relationships Between the Hypersensitive Reaction and Higher Animal Immune Systems
IV. Gene-for-gene Complementarity
V. Recognition Mechanisms in Gene-for-Gene Systems
  A. Genetic Constraints to Model Building
  B. Constraints Imposed by the Phytoalexin Expression Mechanism
  C. Mechanisms in Plants for Recognition of Pathogens
  D. Surface Molecules of Plant Pathogens
VI. Fungal Surface Carbohydrates: Their Possible Role as Specific Recognition Elements in Gene-for-Gene Systems
VII. The Elicitor Receptor Model
VIII. Additional Predictions and Further Tests of the Model
  Acknowledgements
  References

I. INTRODUCTION

An understanding of the biochemical mechanisms underlying gene-for-gene specificity in plant–parasite systems will have eventual utility to resistance breeding strategy, gene deployment and agronomic cropping practices. I will review some of the genetic and biochemical observations relevant to this objective and will present a current model for recognition and specificity in gene-for-gene systems, the pathogen elicitor–host receptor scheme. Certain predictions based on the model will be advanced and additional tests that have yet to be made will be noted. I shall draw on the better developed research with higher animal immune systems, and will inject liberal speculation, the wisdom of which may be transient. Except where noted, the terminology will follow Robinson (1969) and Russell (1978). I will emphasize mechanisms underlying specific (viz. single gene mediated) defence of plants to pathogens and other pests. Since this has often been confused with mechanisms conferring non-specific or general resistance, I will attempt to distinguish these fundamentally different plant defence strategies at the outset.

II. REQUIREMENTS OF SUCCESSFUL PATHOGENS

Disease defence in plants is based on an array of distinct and sometimes subtle mechanisms. In order to colonize host tissue and reproduce, a successful parasite must have accumulated the genetic information to eliminate, overcome, avoid or escape all of the host defences encountered. I liken the interactions involved to a lock and key analogy where the notches on the parasite "key" (Fig. 1) are those factors required to allow colonization of plant tissue and/or to overcome host resistance factors. Pathogens of different plants must have different keys depending on the host and their relationship with it. Different pathogens on the same plant or on different organs of the same plant may also require unique notch patterns depending on their particular mode of pathogenesis, environmental factors, host age and so on.

A non-specific pathogen such as *Sclerotium rolfsii* has a relatively simple key, with notches for the production of pectic enzymes and oxalic acid (Bateman and Beer, 1965; Kritzman *et al.*, 1977), the enzymes needed for detoxification of or insensitivity to preformed host antibiotic factors (e.g. DeFago, 1982) (Fig. 1a) and a cell surface coated with melanins or glycoproteins that do not elicit active host defence. The notches considered here are concerned with overcoming the general resistance of the host: that is, of establishing what has been called basic compatibility (Ellingboe, 1976). Most, but not necessarily all, of these notches confer positive functions that allow pathogenesis by overcoming static barriers (Royle, 1976) or by killing host cells. Some, however, represent greater complexity. For example, pathogenic strains of *Pseudomonas solanacearum* and *Erwinia amylovora* produce significant amounts of extracellular polysaccharide (Ayers *et al.*, 1979; Sequeira and Graham, 1977). Non-pathogenic strains lacking extracellular polysaccharide are engulfed by plant cell walls, presumably by an initial interaction between their lipopolysaccharide and a lectin on the surface of the host cells. The interpretation is that pathogenic strains produced extracellular

(a) Non-specialized pathogen

General (basic) compatibility

Possible notch functions:

pectic enzymes for maceration of host tissue
toxins for killing host cells and/or destroying permeability barriers
enzymic degradation of preformed toxicants or phytoalexins
production of surface molecules that prevent detection of underlying polymers which would otherwise elicit a hypersensitive response (HR)
production of substances that suppress expression of host HR

(b) Specialized pathogen

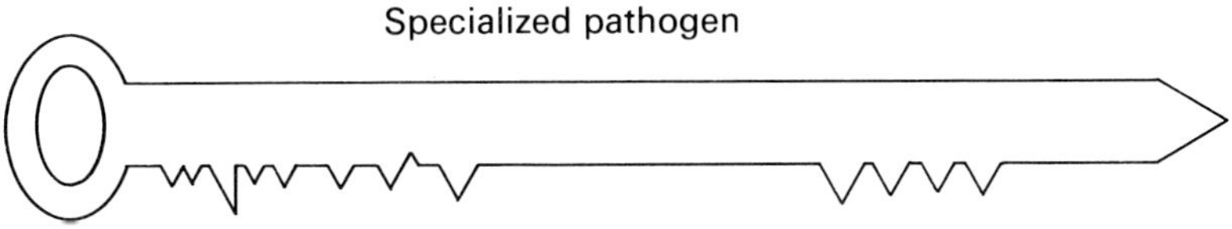

Region of general (basic) compatibility

Possible notch functions:

allowing spores to adhere to the host surface
allowing germ tubes to adhere properly to cuticle and reach stomata or to directly penetrate tissue
penetration of host cell walls by infection hyphae
mechanisms to escape preformed toxicants
formation of haustoria or other specialized structures
may also include any of the notch functions above for non-specialized pathogens

Region of specific compatibility

Possible notch functions:

coding for specific molecules, probably on the cell surface, that may be recognized by the host and invoke active defence.

Fig. 1. Key analogy of the various requirements for parasites to attain success on a certain host plant. Notches on the keys represent factors allowing the neutralization or escape of certain host defence mechanisms. Non-specific pathogens have simpler notch patterns than specialized pathogens.

polysaccharide which masked their lipopolysaccharide, thus preventing recognition of the latter and allowing the bacteria to multiply in the host. In the analogy, extracellular polysaccharide production is therefore a notch on the parasite key.

If the pathogen lacks any of the notches to overcome or escape certain barriers (such as the lectins above), then these constitute defence mechanisms. Such non-host resistance mechanisms (Heath, 1980) are often simple, but a number of them may operate co-operatively or sequentially (e.g. Heath, 1974). For instance, if a small percentage of infecting propagules survive an early host defence mechanism, they may be arrested by later barriers. Examples would be the inability to escape inhibition by preformed host toxins, to obtain sufficient nutrients in the host's intercellular spaces, to produce haustorial mother cells in proper juxtaposition to host cells or to avoid a hypersensitive defence reaction (see below).

The more specialized the host–parasite relationship, the more complex the parasite notch pattern required, the extreme being found in the highly specialized obligate parasites (Fig. 1b). Non-specialized parasites often have wide host ranges and specialized parasites do not. The notch patterns explain this. If the parasite has a very complicated notch pattern, allowing an intimate and complex relationship with its host, the chances of the same key working on another, even closely related, plant are small. Experience from breeding for resistance tells us that resistance to non-specialized pathogens is not common, whereas resistance to highly specialized pathogens is widespread and often simply inherited. This latter phenomenon is specific resistance and involves a distinct region of the parasite key which may not exist on the key of the simple pathogen: the region of specific compatibility (Fig. 1b). To be successful, the parasite must evolve notches in this region that do not activate resistance in the host. They must operate negatively. Unlike other notches that are concerned with establishing basic compatibility with the host (Fig. 1b), success on a plant with resistance gene "x" requires that the pathogen modify the notch corresponding to "x" such that efficient host recognition of the pathogen no longer occurs. I will elaborate on such gene-for-gene complementarity later.

What happens when pathogen specialization is extended to the point of mutualism, as occurs between vesicular–arbuscular mycorrhizas or *Rhizobium* spp. and their hosts? In these cases, the micro-organisms have acquired all necessary notches in both the regions of general and specific compatibility (Fig. 1). Host and parasite have evolved such that specific recognition of the micro-organism does not occur to invoke active host defence. In apparent contradiction, specific recognition occurs between

Rhizobium spp. and compatible legumes to confer compatibility (Dazzo *et al.*, 1978; Schmidt, 1979). This, however, is not an exception to the tenets that I have advanced since mycorrhizas and rhizobia are, in net terms, beneficial to the host and selection pressure has not been placed on the plant to evolve specific (or general) detection of them or to invoke active defence. Rather, the tendency has been to perpetuate mechanisms favouring compatibility.

To summarize, the major points of the key notch analogy are as follows. (i) If the parasite lacks one or more of the necessary notches in the basic compatibility region for a certain plant lock, it will not be successful on that plant, and the host barrier corresponding to the missing notch will be an effective defence mechanism. Notches in this region represent weapons for overcoming structural or chemical barriers as well as other mechanisms facilitating colonization of the host (Fig. 1a). (ii) The more specialized its relationship with the host, the more complex must be the parasite's notch pattern in the region of general compatibility. (iii) The growth rate of specialized pathogens in their hosts is a result of molecular interactions occurring primarily in the portion of the key denoted as the region of specific compatibility (Fig. 1b).

The remainder of this review will emphasize gene-for-gene interactions and the specificity mechanisms underlying them. This is not intended to minimize the significance of general defence mechanisms in nature. Along with general resistance conferred by the hypersensitive reaction, the earlier switching points discussed above and by Heath (1980) are the mechanisms relegating the vast majority of potential parasites to being non-pathogens on a certain plant species. It is those few survivors of evolution which have specialized genetic and biochemical relationships with their hosts that I will consider further.

III. THE HYPERSENSITIVE REACTION (HR)

The hypersensitive reaction (HR) is an active defence system of higher plants typified by necrosis of host cells at the infection site and a reduction in colonization rate of the pathogen. The hypersensitive reaction appears to function as a last line of defence to pathogens because it is only invoked when intimate contact of parasite and host cells occurs. When this happens with infective organs of the pathogen, it is too late for the host to rely on static exclusion mechanisms since these have already failed. Unlike static defence mechanisms, the HR is biochemically induced. Parenchyma and other plant cell types have the potential to respond hypersensitively, providing that recognition of some feature of the invading parasite initiates the mechanism. The recognition systems are bimodal: one level is non-specific, modulating general hypersensitive resistance of a plant species to an entire pathogen species, and is triggered by some of the notches in the basic compatibility region of the parasite key (Fig. 1); the second involves specific hypersensitive resistance, which is initiated by notches in the specific compatibility region. The latter is usually controlled by single resistance genes in the host.

If an incompatible plant genotype recognizes an invader, the HR will reduce its rate of colonization to less than that on a compatible genotype. Selection pressure due to such a host resistance gene may, however, result in the selection of an allele in the parasite that at least partially restores compatibility, thereby giving rise to gene-for-gene complementarity. A major thesis of this paper is that the recognition elements controlling the HR in gene-for-gene systems are race-specific receptor molecules in the host.

Specific hypersensitive resistance is controlled by many single dominant resistance genes in plants, but is the converse also true? Do single dominant resistance genes generally modulate the expression of hypersensitive resistance? Current evidence indicates that they do. Plant reactions such as the necrotic and chlorotic flecking observed after inoculation of leaves with incompatible races of fungus pathogens exemplify classic HR (Stakman, 1915; Ward, 1902). Some workers, however, believe that single gene resistance which does not lead to visible flecks may not involve an HR. Likewise, single dominant genes leading to reduced sporulation of the fungus, slower lesion enlargement and/or a longer interval between inoculation and first visible symptoms (latent period) may be expressed by mechanisms distinct from the classic HR. However, there is currently no evidence to support either contention.

In the case of rapid restriction of parasite development, two genes in wheat against powdery mildew have been well studied. Slesinski and Ellingboe (1969) observed that the *Pm1* and *Pm4* genes restricted fungus development at 24–28 h and 20–24 h, respectively. Since the majority of infecting fungus propagules were inhibited even before forming haustoria, it was speculated that the mechanism(s) for inhibition of growth might be constitutive and not inducible. However, Slesinski and Ellingboe (1969), Hyde and Colhoun (1975) and Haywood and Ellingboe (1979) observed the appearance of distorted, collapsed and necrotic host cells surrounding the incompatible penetration sites. Thus, the observations support the association of classic HR with expression of the rapid-acting *Pm1* and *Pm4* genes. The same can be said for similar genes in other host–parasite systems (e.g. Harder *et al.*, 1979; Littlefield, 1973), but in no gene-for-gene system have such genes been shown conclusively not to function by conferring HR.

Resistance genes conferring small effects on the restriction of pathogen colonization also appear to function through the HR. The L^7 and *K* genes in flax allow substantial vegetative development and some sporulation of incompatible races of *Melampsora lini*; however, they appear to be expressed in the same, albeit slower, manner as faster-acting genes, and lead to hypersensitive host cells and the accumulation of phytoalexins (Coffey, 1976; Keen and Littlefield, 1979; Littlefield, 1973). Similar gene expression has also been reported in other host–parasite systems (Lazarovits and Higgins, 1976; Mares, 1979).

Despite genetic (Ellingboe, 1978) and biochemical (Keen and Bruegger, 1977) evidence to the contrary, induced susceptibility or induced accessibility models have been advanced as alternatives to the idea that hypersensitive resistance is active resistance (Daly, 1972; Ouchi *et al.*, 1974). Induced susceptibility implies that resistance is a passive plant response and that compatibility is the active and specific event in host–parasite interactions. The work of Ouchi *et al.* (1976) with the *Erysiphe graminis* f.sp. *hordei*-barley system provides convincing evidence that inoculation of leaves with a compatible mildew race allows subsequent colonization of the same leaves by an incompatible race. However, none of the evidence proves that the first inoculation had in fact induced accessibility by causing a metabolic change in host cells required for compatibility to the normally incompatible race. For a process to be considered biochemically inducible, a series of *de novo* anabolic events, usually including gene transcription, must take place. The simplest way to detect such *de novo* events is with specific chemical agents that affect host but not parasite metabolism. An inducible process should be blocked if the proper inhibitors are used. I am unaware of agents that convert a normally compatible host reaction to incompatible without exhibiting a direct inhibitory effect on the pathogen or invoking active defence. On the other hand, exogenous inhibitors of HR expression have been frequently reported (Keen *et al.*, 1981; Oku *et al.*, 1975a; Tani and Yamamoto, 1979; Yoshikawa *et al.*, 1978a, b), and the observed conversion of normally incompatible host reactions to compatible is one of the factors indicating that hypersensitive resistance is indeed an inducible process.

The alternative interpretation of Ouchi's experiments on accessibility is that infection by the compatible mildew race had altered host cell metabolism in such a way that the HR could not occur when those cells were confronted with later invasion by the incompatible race. Therefore, susceptibility or accessibility may not have been induced, but resistance suppressed, as the results of Oku *et al.* (1975a) directly indicate.

The evidence indicating that host-specific toxins may induce susceptibility is subject to the same error of interpretation. Parasites producing such toxins are only compatible on certain host cultivars (Scheffer, 1976), and the latter frequently contain a single allele for both sensitivity to the toxin and susceptibility to disease. Toxin insensitive cultivars are often claimed to be passively resistant to the pathogen, but are probably undergoing a classic HR defence reaction, as indicated by the work of Paddock (1953) with *Helminthosporium victoriae*. Accordingly, the role of the selective toxins is most likely to suppress the HR in toxin-sensitive plant genotypes. It is alternatively possible that such toxins may invoke a massive HR, but the pathogen can withstand the defene reaction.

A. The Hypersensitive Reaction and Phytoalexins

Although the quintessential feature of hypersensitive resistance is restriction of parasite development, we do not completely understand how this occurs. The host cell death classically associated with the HR is belived by some to stop parasite development *per se* by failing to supply nutrients, but this is now generally considered untrue even for obligate plant pathogens (Mayama *et al.*, 1975; Prusky *et al.*, 1980). Nutritional hypotheses are also inconsistent with the findings that the HR is an active, energy requiring process. Other explanations for the HR such as cataclysmic alteration of host cell permeability or variable levels of preformed phenols in the host can also be discarded as general mechanisms since they are not consistent with the experimental evidence.

It is possible that more than one biochemical mechanism is responsible for the restricted pathogen development associated with the HR of various higher plant species, but research with several plants strongly supports a role for phytoalexins and perhaps the related mechanism of induced structural barriers (Vance *et al.*, 1980). Research on phytoalexins has been thoroughly reviewed in the last few years (Cruickshank, 1980; Friend, 1979; Keen, 1981, 1982; VanEtten and Pueppke, 1976). Phytoalexins are generally defined as low molecular weight, antimicrobial products of secondary plant biosynthesis that are induced to accumulate to inhibitory levels in the localized pathogen infection court (Paxton, 1980). They occur at relatively low or undetectable levels in healthy plant tissues, but accumulate in response to inoculation with incompatible pathogens or following treatment with various elicitors

(Keen and Bruegger, 1977; West, 1981), some of which are obtained from pathogens. Once production is initiated at the infection site, phytoalexin accumulation may be extremely high, reaching 10% or more of the tissue dry weight. Such levels greatly exceed those at which most plant pathogens are sensitive.

Perhaps as a consequence of the presence of phytoalexin-degrading enzymes in living plant cells (see Keen, 1981), phytoalexins accumulate preferentially in the dead host cells associated with the HR. It is therefore possible that the dead cells function as reservoirs for phytoalexin accumulation, but may not be directly involved in restriction of parasite development.

B. Relationships Between the Hypersensitive Reaction and Higher Animal Immune Systems

The HR/phytoalexin defence system in higher plants is the functional equivalent of the immune systems in higher animals. The major differences are that the immune systems are frequently systemic while HR is local, the immune systems are generally not self-destructive while the HR is, the immune systems persist but plant HR defence does not, and the effector systems (e.g. immunoglobulins) are highly specific in animals but non-specific in plants (phytoalexins). Despite these differences, there is much in common between the pathogen recognition systems of higher plants and animals.

Immunologists have traditionally been concerned with immunoglobulin-mediated defence, but recently they have directed attention to a distinct, but probably equally significant, inducible defence system, the major histocompatibility complex (MHC) (Klein, 1979; Snell *et al.*, 1976). There are several parallels between the MHC and the higher plant HR (Table I).

The MHC appears to be important for maintenance of the self-integrity of the individual and was discovered because of its role in tissue graft rejection. Certain pollen-style and graft incompatibilities may also be a type of HR in plants (Keen, 1981). The plant and animal defence systems are also both genetically polymorphic, exhibiting the phenomena of closely linked genes and allelism.

There are distinct recognition and effector systems in both the HR and MHC. Information exchange in both systems (the determinative phase) has been shown to be an initial step, temporally isolated from the events that actually restrict pathogen development (the expressive phase) (Keen and Bruegger, 1977). Both systems are also similar in that they rapidly recognize a large array of macromolecular foreign elements

Table I. Similarities betwen the higher plant hypersensitive reaction-phytoalexin system and the vertebrate immunoglobulin-major histocompatibility complex systems.

(1) Both are important in non-self recognition; viz. maintenance of the integrity of the individual.
(2) Both are controlled by complex, polymorphic genetic systems, including closely linked genes, allelic genes and possibly split genes.
(3) Both have temporally and functionally distinct recognition and effector systems.
(4) Both systems have constitutive elements that efficiently recognize certain features of incompatible pathogens with a high degree of specificity, suggesting the occurrence of unique cell surface receptors.
(5) The effector mechanisms in both are inducible, involving active, energy requiring processes.

in the environment with a high degree of specificity. This suggests that both code for specific cell-surface receptors, a contention supported by only indirect evidence in plants (Ellingboe, 1976; Keen, 1982) and animals (Janeway, 1980; Klein, 1979).

Finally, the expressive or effector phases for both the HR and the MHC systems appear to be active, energy requiring processes requiring nucleic acid transcription and translation. The MHC expressive phase in complex and not fully understood, while the accumulation of phytoalexins is a major aspect of the higher plant HR. The inducible nature of the two systems renders them susceptible to blockage by various drugs and metabolic inhibitors, sometimes called suppressors.

The biochemistry of recognition in the MHC is better understood than for the HR. However, the similarities noted in Table I for the two defence systems indicate that the research with animals should have considerable impact on the formulation of recognition models in plants.

IV. GENE-FOR-GENE COMPLEMENTARITY

The quadratic check (Fig. 2a) describes the main genetic features of the gene-for-gene relationship for single controlling loci in plant and parasite (Day, 1974; Flor, 1971; Sidhu, 1975). A more important aspect of gene-for-gene complementarity is the reciprocal check (Fig. 2b). This feature makes these interactions distinct from pathogen recognition

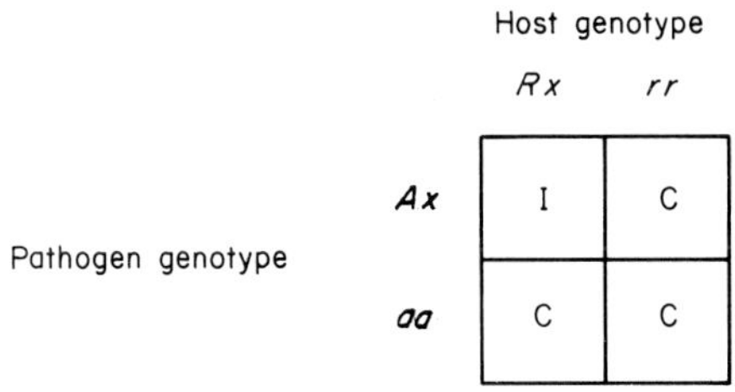

Fig. 2a

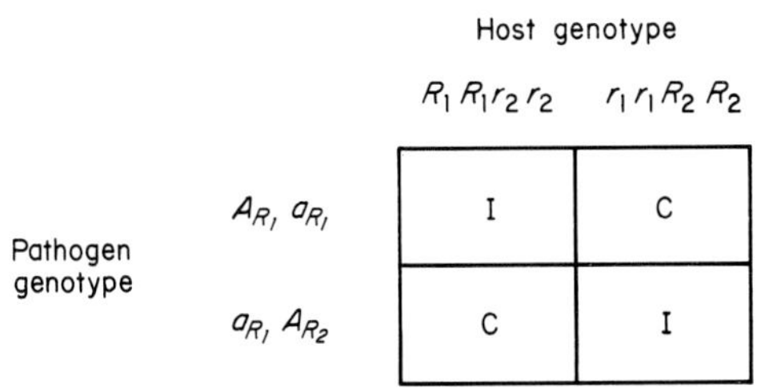

Fig. 2b

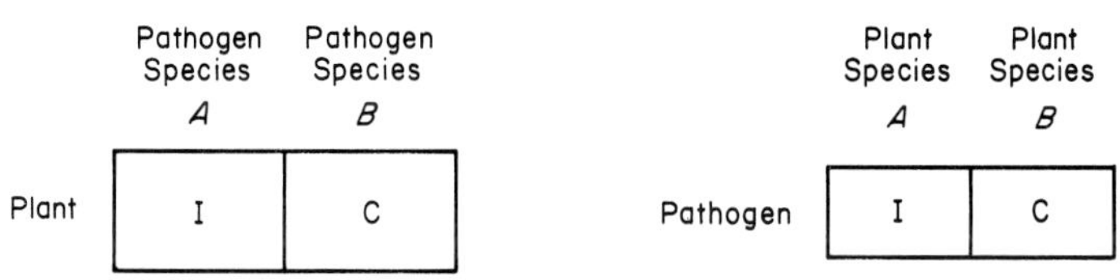

Fig. 2c

Fig. 2. The quadratic (2a) and reciprocal (2b) checks, diagnostic of gene-for-gene systems. The simple specificities shown in (2c) describe systems based on general resistance mechanisms. I = incompatible or resistant; C = compatible or susceptible.

mechanisms that are part of the general resistance system and which exhibit only simple plus/minus specificities (Fig. 2c). Relatively few host–parasite systems have been sufficiently well studied genetically to rigorously prove gene-for-gene complementarity (Day, 1974). However, many others behave as gene-for-gene interactions according to the natural appearance of virulent parasite races predicted by the prevailing resistance genes in the host population (Day, 1974; Sidhu, 1975). Host–parasite systems exhibiting reciprocal cultivar-race characteristics

(Fig. 2b) may therefore be considered gene-for-gene even in the absence of formal genetic studies in the parasite or host.

Nelson (1979) has outlined the probable evolution of gene-for-gene systems. Since pathogens frequently evolve to virulent races when confronted with resistant plant cultivars, one could question the evolutionary legitimacy of resistance genes. What are the selective advantages of these genes that maintain their evolutionary presence in plant genomes? We know of at least four possible factors. (1) If plants of mixed resistance genotypes emerge from dormancy and are confronted with pathogen inoculum of mixed virulence genotypes, then the prevailing plant resistance genes will delay the rate of disease development by reducing the effective initial inoculum (Robinson, 1980; Wolfe and Barrett, 1981). (2) Although the conversion from dominant to recessive virulence alleles does not destroy basic life processes in pathogens, it may decrease the biological fitness of the organism (Leonard and Czochor, 1980; Van der Plank, 1975); this stabilizing selection suggests that, in the absence of a complementary resistance gene in the host, avirulence gene products confer a selective advantage and the parasite may pay an epidemiological price for possessing the complementary virulence allele. (3) Some resistance genes cause only partial restriction of the parasite infection rate; unlike genes conferring near immunity, substantial development of an incompatible parasite race occurs in hosts carrying these genes, and extreme selection pressure is therefore not placed on the pathogen population for selection of virulent genotypes, especially if the resistance genes are pyramided (Parlevliet, 1979). (4) Growth of a virulent race of *Erysiphe graminis* f.sp. *tritici* was less on wheat leaves carrying the *Pm1* resistance gene than on *pm1* leaves (Slesinski and Ellingboe, 1971).

These results indicated that full compatibility was not restored when the fungus carrying the *p1* virulence allele was grown on *Pm1* wheat as compared to *pm1*. Similar findings were later made with mildew strains that were virulent on wheat plants carrying the *Pm4* resistance gene (Martin and Ellingboe, 1976). Three virulent isolates of the fungus (all presumed to carry *p4* virulence alleles) eventually formed apparently fully compatible, sporulating colonies on the *Pm4* near-isoline, but none was in fact fully compatible relative to development on the *pm4* line when sensitive measurement techniques were used. One virulent race formed secondary elongating hyphae more slowly on *Pm4* than on *pm4* wheat, but the same percent of successful infections was eventually attained. With two other virulent *p4* strains, only 19% and 33% of infecting conidia formed normal elongating secondary hyphae after 28 h. The evidence therefore indicated that the *p4* virulence genes in the fungus allowed the regaining of some but not all compatibility on the *Pm4* host genotype. This means that the three compatible combinations in a standard quadratic check (Fig. 2a) are not necessarily equivalent (see Ellingboe, 1980) and plant resistance gene function may not be entirely abrogated when confronted with a virulent parasite race. This conclusion has recently been confirmed and extended by Nass *et al.* (1981) who incorporated four different mildew resistance genes into a single wheat line and observed greatly reduced infection efficiency, colonization and sporulation by a mildew race containing virulence alleles for all of the resistance genes. Similar observations to those above were also made for the L^1 gene of flax to compatible race 1 of *Melampsora lini* (Keen and Littlefield, 1979). It is likely that many similar cases of ghost resistance genes occur in nature and have not been recognized because the reduction in infection efficiency or infection type of the virulent parasite race by the resistance gene is small and difficult to quantify, especially if only crude visual ratings are made of parasite colonization and the environment is not controlled (see Ellingboe, 1978).

Consider what happens if several genes such as the wheat mildew resistance alleles are incorporated into a single host genome and parasite virulence alleles exist for all of them. Assuming a cumulative effect of the several ghost resistance genes on restriction of the parasite infection efficiency or infection rate, as has been experimentally observed by Nass *et al.* (1981), then the result is precisely what has been called horizontal resistance. As Riley (1973) was aware, such resistance would be controlled by several genes, would cause partial resistance but not near-immunity, and should not be overcome by new virulent parasite races since the prevailing races are already virulent.

The above factors explain the persistence of single disease resistance genes in plant populations. More importantly, however, they offer important clues concerning gene function, to be considered later.

V. RECOGNITION MECHANISMS IN GENE-FOR-GENE SYSTEMS

A. Genetic Constraints to Model Building

Much is known about the behaviour of genes controlling plant disease resistance and about those that control the expression of virulence in pathogens. In addition, much is understood about the nature of phytoalexin production in the HR. One of the voids in plant pathology, however, is the area between these two, the specific recognition events that are the consequences of gene function and which predicate the ultimate expression of resistance. Since knowledge is deficient, the sensible approach is to marshal the available genetic and biochemical information and then propose a testable model that is consistent with the concepts of recognition specificity in other biological systems. Table II summarizes the available genetic information that is relevant to the construction of a recognition model.

Higher plant disease resistance genes are usually inherited as dominant characters, but a few (viz. *sr17* in wheat to stem rust, and *pg 8*, *9*, *12* and *13* in oats to stem rust (Green, 1970)) are inherited as recessives. Genes for virulence in pathogens are generally inherited as recessive characters, but a few exceptions exist (e.g. dominant virulence genes in *Puccinia recondita* complementary to *LrB* and *Lr14B* in wheat (Samborski and Dyck, 1976)). The simplest functional interpretation of the genetic observations is that incompatibility is the specifically determined event in gene-for-gene systems and is conferred by the active gene products of the respective dominant alleles (Ellingboe, 1978, 1981). Thus, compatibility is a passive process relative to incompatibility, and only occurs when positive recognition of a pathogen "notch" in the region of specific compatibility does not occur (Fig. 1).

Different single resistance genes in the same plant species frequently exhibit distinct restrictive effects on the colonization rate of the pathogen (e.g. resistance genes to wheat stem rust (McIntosh, 1976)). These genes may be referred to as fast or slow, with various intermediates. This important phenomenon suggests that the elements controlled by the various resistance genes have unique recognition efficiencies.

Gene dosage effects (incomplete dominance) have been noted, for instance with the H^t gene in corn (Dunn and Namm, 1970; Calub *et al.*, 1973), and the *Lr9* gene in wheat (Samborski, 1963). In these cases, the heterozygote exhibits intermediate resistance expression between the homozygous recessive and the homozygous dominant. Similarly, McIntosh (1976) showed that the *Sr15* and *Sr22* resistance genes in wheat gave additive expression of resistance to incompatible stem rust races. Baker (1966) observed the same for two resistance genes in oats to crown rust. It has also been recognized that the genetic background of a plant cultivar may affect expression of single resistance genes (viz. *Sr5* and *Sr6* in wheat (Rohringer *et al.*, 1979)). These effects are probably due to the presence or absence of additional resistance genes with small effects on pathogen infection type in certain genetic backgrounds (viz. Dyck and Samborski, 1974; Law and Johnson, 1967; Parlevliet, 1976).

Table II. Genetic factors associated with gene-for-gene systems.

(1) Resistance genes in plants and avirulence genes in fungus pathogens are, with a few exceptions, generally inherited as dominant characters.
(2) The timing of resistance gene expression is variable. This may range from fast to slow invocation of defence and restriction of pathogen development.
(3) Additive gene dosage effects occur and epistasis operates with resistance alleles. Additive effects of resistance genes, each giving partial reduction in pathogen infection type, are also known.
(4) Resistance gene expression may vary with the genetic background of the host cultivar.
(5) Horizontal resistance is functionally the same as vertical or single gene resistance.
(6) Resistance gene function is probably constitutive; resistance genes are probably structural in nature and the primary products function as specific recognition elements.
(7) Batteries of closely linked R genes exist in some plants and true allelism also occurs.
(8) Some resistance genes are temperature sensitive.
(9) There are a few cases where the same resistance gene is effective against avirulent races of different pathogen species.
(10) Stabilizing selection appears to operate in certain gene-for-gene systems.
(11) Resistance gene function may not be entirely abrogated when the plant is confronted with a virulent pathogen race.
(12) Certain resistance genes lead to mesothetic reactions.
(13) Single gene resistance may be more durable (permanent) against bacteria and viruses than against fungi.
(14) Virulence genes are generally not tightly clustered as in the case of certain plant resistance genes, and no examples of allelism are known.
(15) In a few cases, two avirulence genes in the parasite complement one resistance gene in the host.

Recent research has strongly indicated that horizontal resistance, *sensu* Van der Plank (1963), is functionally one or a pyramid of additive monogenes, each leading to variable but relatively small effects on restriction of the parasite colonization rate (Nelson *et al.*, 1970; Parlevliet, 1978; and see Section IV). As Ellingboe (1981) states,

> The results over the last 10 years have led to the suggestions that horizontal resistance is that resistance which has not yet been shown to be vertical.

One of the major features of horizontal resistance, its permanancy or durability to emergence of virulent parasite races (and the corollary that single gene resistance is not durable), is not supported by experimental evidence. Some durable resistance originally thought to be horizontal has turned out to be single gene mediated (viz. the *St-1* gene in rice (see Parlevliet, 1979)), and several well known single resistance genes have remained durable for many years (e.g. the *T* gene in barley against *Puccinia graminis tritici*, resistance of cucumbers to *Cladosporium cucumerinum*, resistance in cabbage to *Fusarium* yellows, the *N* gene in tobacco for resistance to TMV, the *Sr26* gene in wheat and resistance to common mosaic virus in bean). All these observations demonstrate that single resistance genes have widely different phenotypes and differentially affect the evolution of virulent parasite races. Biochemical models for recognition of incompatible parasite races must accordingly have the flexibility to account for these findings.

Many plants contain closely linked resistance genes (e.g. the *Rp1* genes in corn (Saxena and Hooker, 1968) and the *M* genes in flax (Mayo and Shepherd, 1980)). In addition, true allelism appears to operate at some resistance gene loci (e.g. the *L* locus in flax (Shepherd and Mayo, 1972), the *Sr7* and *Sr9* loci in wheat and the *Lr2* locus in wheat (Dyck and Samborski, 1974)). These observations have important implications for the structure of the pathogen recognition systems controlled by the various resistance genes.

Expression of some, but not all, resistance genes is also affected by temperature. As temperature increases, resistance to *P. recondita* conferred by the wheat *Lr20* gene decreases (Jones and Deverall, 1977). The *Sr15* and *Sr6* genes are also less effective against wheat stem rust as temperature increases, but the opposite is true of *Sr13*. In addition, plant age markedly affects the expression of certain resistance genes (e.g. *Phytophthora infestans* resistance in tomato (Dowley *et al.*, 1975); the *Lr2* and the *Sr2* genes in wheat (Sunderwirth and Roelfs, 1980)). These observations suggest that the recognition factors conferred by resistance genes may be heat labile and are produced in different amounts as a function of cell age and differentiation.

There is now good reason to think that certain single resistance genes confer resistance to taxonomically different pathogens (viz. the *Sr24/Lr24*, *Sr15/Lr20*, *Sr31/Lr26*, and *Sr25/Lr19* genes in wheat for resistance to *P. graminis tritici* and *Puccinia recondita*, respectively (McIntosh, 1977; Roelfs, personal communication), and a single gene in soybeans for resistance to *Heterodera glycines* and *Rotylenchulus reniformis* (Rebois *et al.*, 1970)). The apparent existence of independent genes conferring recognition of different pathogen species suggests that the different avirulent pathogens may contain functionally similar recognition factors.

Stabilizing selection shows that certain virulence alleles in parasites affect biological "fitness" such that they are selected against when the complementary resistance gene is removed from the prevailing host population (Crill, 1977; Leonard and Czochor, 1980; Person *et al.*, 1976). Thus, it may render unnecessary virulence genes rare in the pathogen population. Although the extent of stabilizing selection is still in debate, in cases where it operates, the avirulent pathogen genotype must have a survival advantage over the virulent genotype. This has implications for the nature of avirulence gene products.

An outgrowth of stabilizing selection is the concept of strong and weak resistance genes (van der Plank, 1975). Weak genes are those that do not result in significant loss of fitness of the parasite containing virulence alleles, while strong genes or combinations often do. For instance, van der Plank (1975) noted that the *Sr9d/Sr6* combination of genes for resistance to stem rust in North American wheat behaves as strong genes. Some of the controversy relating to the occurrence of stabilizing selection may result from variation in environment or other reasons (Leonard and Czochor, 1980). For instance, Martens *et al.* (1977) noted that the *Pg2/Pg4* resistance gene combination for resistance in oats to *P. graminis* f.sp. *avenae* behaved as strong genes in New Zealand but as weak genes in Australia.

Martin and Ellingboe (1976) have shown that virulence genes in three *Erysiphe graminis* f.sp. *triciti* strains complementary to the *Pm4* resistance gene in wheat did not lead to full re acquisition of compatibility (see Section IV). This must be accounted for in any model of gene-for-gene specificity.

Yet another phenomenon to be considered is that certain resistance genes lead to a mixture of infection types following inoculation with a pure strain of a pathogen (viz. mesothetic reactions in wheat to stem rust and the X reaction to mildews (Ellingboe, 1978)). Equally interesting are the Z and Y reactions of certain corn and wheat genotypes to *Puccinia sorghi* and *P. recondita*, respectively, in which proximal and distal portions of leaves exhibit large differences in infection type (see Hooker, 1967). Although nothing is known about the mechanisms underlying these phenomena, recognition models should have the flexibility to account for them.

It appears to me that resistance genes against bacterial and viral pathogens are less frequently overcome by virulent pathogen races than are genes effective against fungi. Although resistance genes have been used less extensively against the former pathogens, and epidemiological factors may contribute differentially to resistance gene durability, it is noteworthy that relatively few gene-for-gene relationships are recognized with bacteria and viruses. Compare this with the case of resistance to nematodes and sedentary insects. Relatively little resistance breeding has been done here either, and nematodes are not disseminated as efficiently as bacterial pathogens of aerial plant parts or vector-borne viruses. However, there are several gene-for-gene relationships with nematode systems (e.g. potato-*Heterodera rostochiensis* (Jones and Parrott, 1965); soybean-*Heterodera glycines* (Triantaphyllou, 1975)) and insect–plant systems (Gallun and Khush, 1980; Tingey and Singh, 1980). The argument is strengthened by the observation that the few recognized cases of parasite specialization with bacteria are all with Gram-negative species. Resistance against Gram-positive bacteria is apparently rarer, but more durable (see Russell, 1978). Gram-negative and Gram-positive bacteria have very different cell surface architecture, which provides a clue concerning recognition systems. I conclude that the differences in durability of resistance genes to various plant pathogens means that recognition molecules in higher plants distinguish an array of pathogen chemicals, and that the propensity of pathogens to alter the chemical features recognized by resistance gene products is variable.

Pathogen virulence genes do not generally appear to occur in tightly linked clusters as is frequently the case with plant resistance genes, although linkage is occasionally reported (Day, 1974; Lawrence *et al.*, 1981). Although virulence genes are usually inherited as recessive characters, a few are inherited as dominant, and some are incompletely dominant (Samborski, 1963). In addition, there is evidence that additional loci may modify the expression of avirulence genes, as indicated by the work of Green and McKenzie (1967), Samborski and Dyck (1968) and Lawrence *et al.* (1981). Although the generality of such inhibitor or modifier genes is not yet clear, their occurrence must be taken into account. The genetic studies to date with parasite virulence genes indicate that their primary gene products must have distinctly different functions from those of plant resistance genes.

There are indications that exceptions to the classic one gene-for-one gene relationship of Flor may exist. For instance, single recessive virulence genes in races 1 and 15 of *Puccinia recondita* conferred virulence on all of the tested *Lr2* resistance alleles in wheat (Samborski and Dyck, 1968). However, independent virulence genes occurred in the same fungus to two alleles of *Lr14* (Dyck and Samborski, 1970). Virulence genes *v4* and *v5* of *Ustilago hordei* independently conferred virulence on the barley *cv.* Himalaya (Ebba and Person, 1975). The pathogen was avirulent only if both of the dominant alleles *V4* and *V5* were present.

Several other cases where two virulence genes complement a single resistance gene are known (Statler, 1977, 1979b, for *Puccinia recondita* on wheat; Statler, 1979a for *Melampsora lini* on flax; Christ and Groth, 1979, for *Uromyces phaseoli* on bean). At first inspection, such two-for-one interactions would seem to be serious exceptions to the gene-for-gene relationship, but I think not. The central feature of the hypothesis, the specific complementarity of host and parasite genes, remains intact. However, occurrence of two-for-one complementarity has implications for the building of recognition models and reinforces the view that virulence gene function must be distinct from resistance gene function.

There appears to have been no attempts to genetically cross isolates of the same pathogen race obtained from different geographic locations in order to ascertain whether their virulence genes are allelic. Based on the above examples of two-for-one complementation, it is possible that certain virulent parasite races may contain independent and possibly unlinked virulence loci complementary to a single host resistance gene. The work of Martin and Ellingboe (1976) with virulent isolates of *E. graminis* f.sp. *tritici* may exemplify this, but genetic studies have not been performed with the fungus isolates to confirm it.

B. Constraints Imposed by the Phytoalexin Expression Mechanism

There is now substantial evidence from several gene-for-gene systems that phytoalexins accumulate to high quantity after inoculation of plants with incompatible but not compatible parasite races (Tables III and IV). The consistent association of high phytoalexin production only with incompatible host reactions in these systems strongly supports their role as mediators of disease resistance. It further points to the existence of specific recognition systems in resistant genotypes, and indicates that these modulate phytoalexin accumulation.

The accumulation of phytoalexins occurs in only a few host cells in the immediate vicinity of the incompatible pathogen race (Keen and Littlefield, 1979; Mansfield *et al.*, 1974; Yoshikawa *et al.*, 1978a, b). Accordingly, the function of the pathogen recognition system in resistant plants must be localized in host cells contacted by the pathogen and those immediately surrounding the infection site.

Phytoalexins exhibit static but usually not lethal antibiotic effects on parasite development *in vitro*. Also, pathogens generally do not die immediately after their development is slowed or arrested in single gene incompatible host reactions (Chakravarti, 1966; Keen and Kennedy, 1974; Littlefield and Aronsen, 1969; Tomiyama, 1955). Thus, even in

Table III. Phenomena associated with phytoalexins and the hypersensitive reaction in gene-for-gene systems

(1) Phytoalexins accumulate only in incompatible reactions, and compatibility results from the passive failure of the plant to accumulate phytoalexins.
(2) Phytoalexin production is strictly localized to the infection site and a few surrounding host cells.
(3) Phytoalexins are generally inhibitory but not lethal to pathogens; parasites involved in gene-for-gene systems do not differentially degrade the phytoalexins of their hosts and are not differentially sensitive to the phytoalexins.
(4) Phytoalexin accumulation is an inducible process and can be blocked by certain heat treatments and specific metabolic inhibitors.
(5) A temperature-sensitive resistance gene in soybean modulates phytoalexin production.
(6) Phytoalexin production may account for the cross protection phenomenon.

rapid incompatible reactions, the pathogen is often arrested rather than killed immediately. With the possible exception of *Colletotrichum lindemuthianum* (van den Heuvel and Glazener, 1975), plant pathogens involved in gene-for-gene relationships with their hosts do not appear to chemically degrade the phytoalexins of their hosts, and various races appear uniformly sensitive to host phytoalexins (Keen and Littlefield, 1979; Oku *et al.*, 1975b). These observations suggest that the success of compatible parasite races is due to their ability to avoid recognition mechanisms that would otherwise invoke the phytoalexin mechanism.

When plant tissues are inoculated with an incompatible pathogen race, phytoalexin accumulation does not occur until several hours later. This lag period results from the requirement for *de novo* synthesis of one or more enzymes in the phytoalexin biosynthetic pathway (Sitton and West, 1975; Zähringer *et al.*, 1979). Consistent with this finding is the observation that metabolic inhibitors block both phytoalexin production and the expression of resistance if applied within a few (5–10) hours following inoculation. If the inhibitors are applied after this period, however, subsequent phytoalexin production and resistance expression occur normally (Yoshikawa *et al.*, 1978a, b). These observations demonstrate that recognition of the incompatible pathogen race occurs rapidly after inoculation but that the biochemical events directly

Table IV. Gene-for-gene systems in which phytoalexin production has been shown to be substantially higher in incompatible than in compatible interactions.

Host–parasite system	References
Tomato-*Cladosporium fulvum*	deWit and Flach, 1979
Bean-*Uromyces phaseoli*	Elnaghy and Heitefuss, 1976
Potato-*Phytophthora infestans*	Sato *et al.*, 1971; Varns *et al.*, 1971
Cowpea-*Phytophthora vignae*	Partridge and Keen, 1976
Soybean-*Phytophthora megasperma* f.sp. *glycinea*	Keen, 1971; Yoshikawa *et al.*, 1978a, b
Tobacco-*Phytophthora parasitica* var *nicotianae*	Helgeson *et al.*, 1978; Helgeson, unpublished
Soybean-*Pseudomonas glycinea*	Keen and Kennedy, 1974; Keen, 1978
Bean-*Colletotrichum lindemuthianum*	Bailey *et al.*, 1980; Rahe *et al.*, 1969
Soybean-*Meloidogyne incognita*	Kaplan *et al.*, 1980
Bean-*Pseudomonas phaseolicola*	Gnanamanickam and Patil, 1977; Lyon and Wood, 1975
Flax-*Melampsora lini*	Keen and Littlefield, 1979
Barley-*Erysiphe graminis* f.sp. *hordei*	Oku *et al.*, 1975a, b
Sugarbeet-*Cercospora beticola*	Johnson *et al.*, 1976
Oat-*Puccinia coronata* var. *avenae*	Mayama *et al.*, 1981
Coffee-*Hemileia vastatrix*	Rodrigues *et al.*, 1975
Corn-*Helminthosporium turcicum*	Lim *et al.*, 1970

leading to cessation of pathogen development (viz. phytoalexin biosynthesis) occur subsequently and irreversibly.

The use of elevated incubation temperatures (Gabriel *et al.*, 1979; Mayama *et al.*, 1975) is a valuable method for converting certain single gene mediated resistant reactions to compatible ones because it does not involve the use of chemical inhibitors that may have unknown side-effects on host metabolism. Temperature elevation has recently been shown to reverse resistance expression and phytoalexin production (Keen *et al.*, 1981). Hypersensitive host cell death and glyceollin accumulation occurred when leaves of the soybean *cv.* Harosoy inoculated with incompatible race 1 of *Pseudomonas glycinea* and incubated at 22°C. However, neither response was observed when the inoculated plants were maintained at 31°C and bacterial populations were about 50-fold greater at the higher temperature. This suggests that resistance expression was temperature sensitive and that phytoalexin production has a role in restriction of bacterial populations. The experiments also have considerable significance for the specific recognition mechanism of the incompatible race 1 bacteria, since other incompatible races of *P. glycinea* caused hypersensitive host reactions and glyceollin accumulation at both 22°C and 31°C, and race 1 led to HR and phytoalexin production on the soybean cultivar "Chippewa" at both temperatures (Keen *et al.*, 1981). The observations indicate that elevated temperature probably interferes with production or stability of a recognitional element involved in the interaction of the race 1 bacteria with *cv.* Harosoy but not *cv.* Chippewa leaves (*cf.* Gabriel *et al.*, 1979).

Phytoalexins have been shown to mediate the cross protection of host tissues against compatible pathogen races when they are inoculated jointly with or following an incompatible race (Keen and Kennedy, 1974; Littlefield, 1969; Paxton and Chamberlain, 1967). This indicates that the incompatible race is physiologically dominant in directing the plant response. This is an important point because it establishes that the observed phytoalexin production and plant infection type are consistent with the genetic dominance of resistance and avirulence alleles.

C. Mechanisms in Plants for Recognition of Pathogens

Although little is known about the biochemistry of pathogen recognition in gene-for-gene systems, recent information and interpretations have led to the idea that recognition is frequently modulated by surface molecules of host and pathogen. Most of the evidence has come from recognition systems involved in non-specific resistance (e.g. Goodman *et al.*, 1976; Klement, 1977; Kojima and Uritani, 1974; Sequeira, 1979). These systems, however, have only plus–minus specificities (Fig. 2c), in which some pathogen strains are pathogenic and some are not, unlike the reciprocal specificities typical of gene-for-gene systems (Fig. 2b). However, with one gene-for-gene system, Stall and Cook (1979) obtained evidence indicating that physical contact of an incompatible race of *Xanthomonas vesicatoria* with pepper cells was necessary for resistance expression. Bruegger and Keen (1979) also observed that cell envelope preparations from incompatible, but not compatible, races of *Pseudomonas glycinea* specifically elicited production of glyceollin in soybean. These observations may indicate that certain molecules associated with the cell surface of the incompatible bacteria are specifically recognized by host cells. They are also analogous to data obtained with surface glycoproteins of fungal pathogens, to be discussed later.

The surface location of recognition molecules is also implied from the findings discussed in the previous section indicating that the determinative (recognition) phase is of relatively short duration (one to a few hours). The relatively rapid establishment of incompatible recognition is most readily explained by interaction of cell surface recognition molecules.

D. Surface Molecules of Plant Pathogens

If the suggestions in the preceding section pointing to the importance of surface molecules are correct, then it is important to consider what these are in the major groups of plant pathogens, especially those involved in gene-for-gene relationships with their hosts. Surface molecules are quite different for the various groups of pathogens (Table V), a fact that implies host recognition strategy may differ depending on the pathogen type.

1. Viroids and Viruses

The simplest plant pathogens, the viroids, have no protein capsids but consist only of nucleic acid; perhaps for this reason, there are no known resistance genes against viroids in plants. Viruses, however, have coat proteins or glycoproteins, and many elicit hypersensitive defence reactions in plants. In spite of this and the well-known efficiency of capsid proteins and glycoproteins as antigens (Burns and Allison, 1975), there is no evidence to indicate that capsid components directly elicit the higher plant HR. The actual recognition system is unknown, but may involve plant recognition of replication proteins or processed messenger RNAs coded for by the viral genome.

2. Mycoplasma-like Organisms

There are no known resistance genes against phytopathogenic mycoplasmas or spiroplasmas. This may be due to the fact that these organisms appear to lack cell walls and surface carbohydrates. Instead, the studied organisms possess protein surface molecules. For instance, *Spiroplasma citri* contains a major surface membrane protein, spirallin, thought to be important in membrane integrity (Wroblewski, 1979).

3. Nematodes

Nematodes are bounded by a thickened cuticle with a polyanionic surface (Himmelhoch *et al.*, 1977). *Caenorhabditis* spp. were shown to bind several plant lectins (Zuckerman *et al.*, 1979), but little is known concerning the chemistry of the cuticle surface. Cuticular surface proteins, however, have been shown to differ in various developmental stages of the animal pathogen *Trichinella spiralis* (Philipp *et al.*, 1980).

Table V. Surface molecules of the major groups of plant pathogens and their possible role in defence recognition by higher plant and higher animal cells.

Pathogen group	Surface molecules	Evidence for role in defence recognition: Higher plants	Evidence for role in defence recognition: Higher animals
Viroids	None—only nucleic acid	None	None
Viruses	Capsid proteins or glycoproteins	None	Capsid proteins and glycoproteins as antigens
Mycoplasmas and spiroplasmas	Proteins associated with the cell membrane; no evidence for carbohydrate conjugates	None	Several membrane proteins are antigenic (Kenny, 1975)
Nematodes	Protein-containing thickened cuticles; evidence for glycoproteins is uncertain	?	?
Gram-positive bacteria and certain Streptomycetes	Peptidoglycan-teichoic acid complex	?	Teichoic acids are major surface antigens
Gram-negative bacteria	Outer membrane consisting of peptidoglycan backbone, lipopolysaccharde and proteins with variable amounts of extracellular polysaccharide	Extracellular polysaccharide (Sequeira and Graham, 1977) and lipopolysaccharides, (Whatley *et al.*, 1980)	Major antigens of serotypes are the O-sidechains of lipopolysaccharides; outer membrane proteins also are antigenic (Jann and Westphal, 1975)
Fungi	Polysaccharide cell walls coated with surface glycoproteins and possibly impregnated with melanins	Glycoproteins are phytoalexin elicitors (Anderson, 1980; Keen and Legrand, 1980)	Immunodominant antigens are the carbohydrate chains of glycoproteins (Ballou, 1974)

Although single gene mediated hypersensitive resistance against nematode pathogens is common in plants, we do not understand the mechanism conferring specific recognition.

4. Bacteria

Gram-positive bacteria differ fundamentally from Gram-negative forms in possessing a teichoic acid based surface peptidoglycan matrix; the latter group have highly developed outer membranes. Single gene resistance to Gram-positive plant pathogens is rare, but several examples of single gene resistance to Gram-negative bacteria are known. The highly organized, plastic outer membrane of Gram-negative bacteria may be partially or completely overlain with extracellular polysaccharide (EPS). The outer membrane is composed of lipopolysaccharide (LPS), certain proteins, phospholipids and a peptidoglycan backbone. Some of the proteins and the O-antigenic sidechains of the LPS are often exposed at the cell surface. The O-sidechains are of special interest because their carbohydrate structure is highly polymorphic. In *Salmonella* spp., dozens of unique carbohydrate structures have been found to account for the antigenic serotypes of various bacterial strains (Jann and Westphal, 1975).

Occurrence of the LPS O-sidechains at the cell surface makes them prime candidates for a recognition role in plant–bacterial interactions. Crosthwaite and Patil (1978) reported that *Pseudomonas phaseolicola* LPS was an HR elicitor in pepper plants, but Mazzucchi and Pupillo (1976) found that LPS-protein complexes inhibited hypersensitive reactions in tobacco, a finding supported by the results of Graham *et al.*, (1977). Our own research has so far failed to reveal phytoalexin elicitor activity in lipopolysaccharide preparations from *Pseudomonas glycinea* (Bruegger and Keen, 1979). Whatley *et al.* (1980), however, found that smooth strains of *Pseudomonas solanacearum* did not induce HR in tobacco, but that rough strains, deficient in much of the O-antigenic polysaccharide, were effective. The failure to elicit HR responses with isolated LPS preparations may be due to the fact that LPS self-associates *in vitro* and accordingly may not behave as when integrated into the bacterial outer membrane. There is nevertheless some support for the idea that structural variation in the O-antigenic sidechains of LPS may be involved with recognition by incompatible plants (for more details see Keen and Holliday, 1982).

5. Fungi

Lectin studies have indicated that certain structural polymers of the cell wall of fungi may be partially exposed at the cell surface (Barkai-Golan *et al.*, 1978). However, the predominant surface molecules in fungi appear to be glycoproteins. The surface glycoproteins of yeasts, usually called mannans or mannoproteins, are composed of polypeptides highly substituted with carbohydrate sidechains of varying lengths and compositions (Ballou, 1974, 1976). These are somewhat analogous to the O-antigenic sidechains of Gram-negative bacterial LPS. Certain of the cell wall associated mannoproteins may be strictly structural elements, but yeasts also produce enzymatically active mannan glycoproteins such as invertase, β-glucosidase and acid phosphatase that are present in the periplasmic space, intermixed with the structural mannan, or released as extracellular enzymes. It is now established that the yeast mannan glycoproteins coat the underlying structural carbohydrate polymers of the cell wall, and are often the major antigenic determinants and surface molecules of yeasts and other fungi (Ballou, 1974; Nakajima *et al.*, 1977). Further, the structures of the carbohydrate portions of the yeast mannans are highly polymorphic. Structural differences have been observed in the carbohydrate outer chains of mannoproteins from yeasts of different species and, of more interest, from strains of the same species (Ballou, 1974).

Yeasts contain relatively short carbohydrate chains covalently linked to threonine or serine residues of the protein backbone and, in addition, possess large carbohydrate chains attached to asparagine residues. These possess a core region of mannose residues and structurally variable outer chains (Lehle, 1980). The latter are of interest because they contain the major points of serologic specificity. Although the corresponding carbohydrate structures from other fungi, including plant pathogens, have not been studied in the detail of the yeast outer chain carbohydrates, they may have a similar role in specific plant recognition.

The polysaccharide structures of the various yeast surface mannoproteins and extracellular glycoproteins are determined by glycosyl transferase enzymes that add sugar residues in specific linkages to the developing carbohydrate chains. The complement of transferase enzymes produced by a strain of fungus predicates the structure of the outer chain carbohydrates (Ballou, 1974; 1976) that are specifically recognized by receptor systems such as those of the immune systems and those thought to be conferred by plant disease resistance genes. The resulting hypothesis that these enzymes are the primary gene products of pathogen avirulence genes has been offered by Albersheim and Anderson-Prouty (1975) and will be considered below.

VI. FUNGAL SURFACE CARBOHYDRATES: THEIR POSSIBLE ROLE AS SPECIFIC RECOGNITION ELEMENTS IN GENE-FOR-GENE SYSTEMS

Several fractions obtained from plant pathogens initiate the HR-phytoalexin system in higher plant cells. These elicitors (Keen *et al.*, 1972; West, 1981) have been most extensively studied with fungi, and activity has so far generally been associated with carbohydrates. For instance, β-glucans obtained from several fungi by autoclaving cell walls or intact hyphae are efficient phytoalexin elicitors (Ayers *et al.*, 1976; Hümme *et al.*, 1978). However, these do not exhibit race or cultivar specificity for phytoalexin elicitation, and their role in plant–parasite interactions is therefore uncertain (Keen, 1981).

Glycoprotein preparations from cell wall fractions or extracellular culture fluids of several fungi elicit plant phytoalexin production, but some of these also do not exhibit race or cultivar specificity (Ayers *et al.*, 1976; deWit and Roseboom, 1980). This may, however, be attributed to structural artefacts caused by the action of lytic enzymes in fungal cultures or during extraction of cell walls by autoclaving for long periods.

The carbohydrate moieties of surface and/or extracellular glycoproteins of *Phytophthora megasperma* f.sp. *glycinea* may be specifically recognized by incompatible genotypes of soybean (Keen, 1978b; Keen and Legrand, 1980; Wade and Albersheim, 1979). For example, cell wall glycoprotein extracts from fourteen isolates representing nine races of *P. megasperma* f.sp. *glycinea* elicited production of the soybean phytoalexin glyceollin in two near-isogenic soybean lines with the same specificity as the fungus isolates from which they were obtained (Keen and Legrand, 1980). Fractionation on gel filtration and anion exchange columns in the presence of dissociating agents led to preparations that contained predominantly glycoproteins, and gave race and cultivar specificity for elicitor activity. Activity was lost when the preparations were treated with periodate, but was retained following boiling or incubation with pronase. These results indicated that a carbohydrate was the elicitor-active moiety. Since no single glycoprotein was purified to apparent homogeneity, it is not yet proven that the specific elicitor activity was covalently-linked with protein, but may have been non-covalently associated. However, the facts that glyceollin was specifically elicited by the preparations and this phytoalexin has been strongly implicated in resistance expression (Keen and Bruegger, 1977) indicate that the wall-associated carbohydrate elicitors are prime candidates for the factors specifically recognized by incompatible plant cells.

Extracellular glycoprotein preparations from incompatible but not compatible *P. megasperma* f.sp. *glycinea* races also protected soybeans against inoculation with a living compatible fungus race (Wade and Albersheim, 1979). The preparations, however, did not specifically elicit glyceollin production, and their relation to the wall associated glycoproteins is therefore unclear.

The structures of the above surface and extracellular glycoproteins have not yet been investigated in detail, but structural differences occurred in the carbohydrate moieties of the extracellular glycoprotein enzyme invertase from three races of *P. megasperma* f.sp. *glycinea* (Ziegler and Albersheim, 1977). Invertase activity is also associated with the mycelium (cell walls?) of *P. megasperma* f.sp. *glycinea* and could be released with dithiothreitol (West *et al.*, 1980). The activity was associated with five apparent isoenzymes, all of which were glycoproteins containing glucose, mannose and a trace of galactose, but these were different from the extracellular invertases in several respects. The glucose/mannose ratios of the mycelial enzymes from two races of the fungus were very different. These results may have considerable significance since Smith and Ballou (1974) showed that the carbohydrate structures of both the surface mannan glycoprotein and the extracellular invertase of Baker's yeast were similar, thereby indicating that the carbohydrate portions of both were synthesized by common glycosyl transferases. If the same is true of *P. megasperma* f.sp. *glycinea*, then the work of Ziegler and Albersheim and West *et al.* predicts that structural uniqueness occurs in the carbohydrate portions of the surface and extracellular glycoproteins of the various fungus races.

The high molecular weight fraction from culture fluids of three races of *Colletotrichum lindemuthianum* also contained glycoproteins and the preparations exhibited race-specific elicitation of phaseollin in red kidney bean cotyledons (Anderson, 1980). These observations, and the work with *P. megasperma* f.sp. *glycinea* glycoproteins, have been reviewed elsewhere (Keen, 1982) and point to a role for surface carbohydrates in determining specific incompatible responses of plant cells to fungal pathogens.

VII. THE ELICITOR–RECEPTOR MODEL

Although it is possible that more than one biochemical recognition

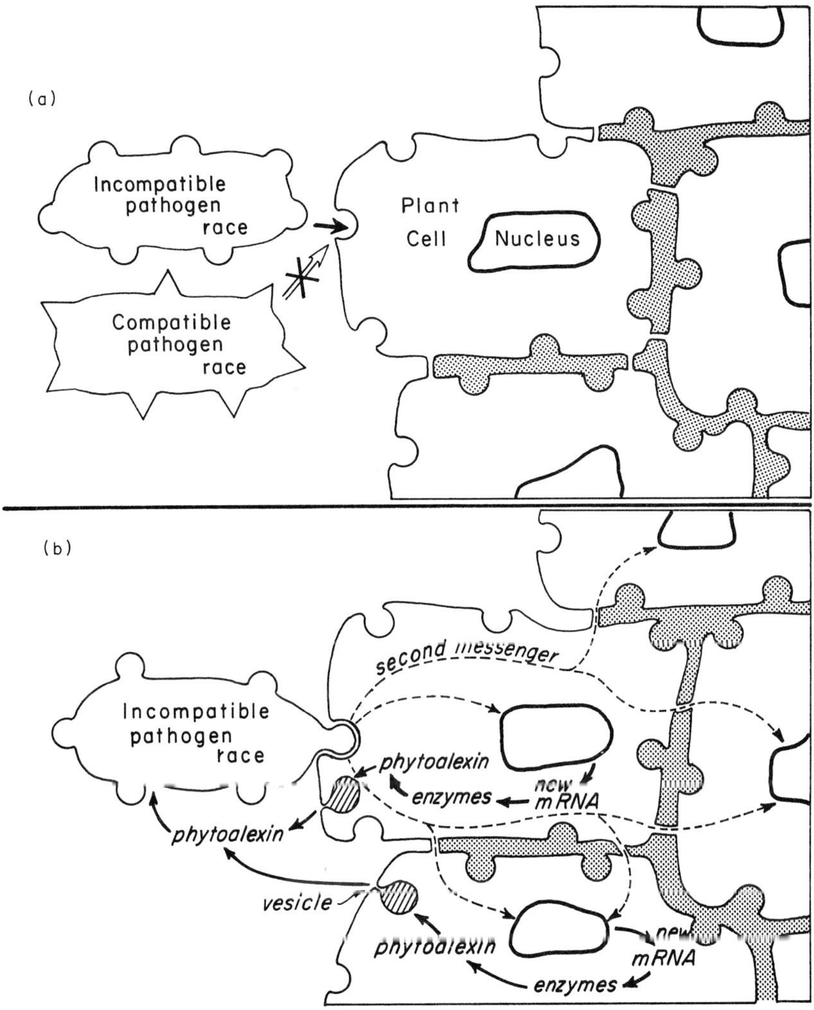

Fig. 3. Schematic representation of the specific elicitor-specific receptor hypothesis for resistance to incompatible parasite races. (a) Compatible races lack surface elicitors that are recognized by the plant receptors and therefore do not initiate active defence. (b) Elicitors from incompatible races interact with the complementary plant cell surface receptors and second messengers transmit this stimulus to the nucleus of the cell contacted and to neighbouring cells via plasmadesmata; the resulting *de novo* transcription and translation leads to production of enzymes in the terminal portion of the phytoalexin biosynthetic pathway and the phytoalexins then formed are secreted via vesicles to the outside of the plant cells where they inhibit growth of the incompatible pathogen race.

mechanism may function in various gene-for-gene systems (see Keen, 1982), the specific elicitor-specific receptor model (Albersheim, *et al.*, 1980; Albersheim and Anderson-Prouty, 1975; Keen and Bruegger, 1977) is most strongly supported by the above evidence and is consistent with most of the genetic and physiological observations listed in Tables II and III. The mechanism, as schematically presented in Fig. 3, entails phytoalexins to account for restricted pathogen development, but other expressive mechanisms such as induced structural barriers could also be modulated by this recognition system in some plants.

A key aspect of the model is the predicted presence in plant cells of specific surface receptors, probably proteins or glycoproteins, that recognize surface molecules of incompatible but not compatible pathogen races. Indirect experiments suggested the occurrence of such receptors (Doke and Tomiyama, 1978; Peters *et al.*, 1978), but no direct demonstrations appeared until recently. Yoshikawa *et al.* (unpublished) have obtained the first experimental evidence suggesting that a biotic phytoalexin elicitor is recognized by a specific receptor on soybean membranes. Radiolabelled mycolaminaran, a branched β-1,3 glucan elicitor produced by *Phytophthora* spp., bound to soybean membrane fractions with a binding constant of 10 μM. The binding was specific, based on competition studies with unlabelled ligand, on the fact that no other carbohydrates competed with binding, and on the correspondence of alterations in both the affinity constant for binding and observed glyceollin elicitor activity of various mycolaminaran derivatives. The results indicate that phytoalexin production is initiated by receptor–elicitor binding and offer precedence for the occurrence of receptors for other pathogen-produced elicitors.

It is most likely that the plant receptors in Fig. 3 occur on the plant cell plasma membrane surface, since it is difficult to envisage them being functional on the cell wall surface. How would wall receptors interact intracellularly? On the other hand, how could pathogen surface elicitors contact receptors on the plasma membrane? One answer may be that many fungus pathogens develop close contact with the host plasma membrane (e.g. Nozue *et al.*, 1979), and this may account for the physical contact of host receptors and fungus surface elicitors. Recent research, however, points to another possibility.

Enzymes from pea and soybean tissues rapidly liberated soluble phytoalexin elicitors from fungus cell walls (Nichols *et al.*, 1980; Yoshikawa *et al.*, 1981). A β-1,3-endoglucanase was purified from soybean cotyledons and liberated high molecular weight carbohydrates containing glucose and mannose from cell walls of P. *megasperma* f.sp. *glycinea* (Keen and Yoshikawa, unpublished). The purified mannan was a potent elicitor of glyceollin, and appeared to give race-specific elicitor activity similar to the glycoprotein elicitors previously discussed. The elicitors liberated from fungus cell walls by the soybean glucanase may therfore be similar to the carbohydrates of the surface glycoproteins. If similar enzyme-mediated liberation of fungus surface elicitors occurs during the host–parasite interaction, this would facilitate contact of the fungus elicitors with the host plasma membrane. The observations would then support the double induction hypothesis, which proposes that plant metabolites effect the release of pathogen elicitors which in turn initiate phytoalexin-mediated defence (see Cruickshank, 1980).

Following interaction of the specific elicitor from an incompatible pathogen race with host cells (Fig. 3), the elicitor–receptor hypothesis requires that a cytoplasmic second messenger, akin to cyclic AMP, phospholipid methylation, or prostaglandin production in other organisms (Hirata and Axelrod, 1980; Kuehl and Egan, 1980), must be formed to transfer information from the plasma membrane receptor to the plant cell nucleus. It is possible that this function may be served by endogenous plant elicitors similar to those observed by Hargreaves and Bailey (1978) and Hahn *et al.* (1981). The logistics of defence also dictate that this messenger must have a degree of systemicity, interacting not only with the nucleus of the host cell(s) initially contacted by the pathogen (and which, together with a variable number of surrounding cells will undergo hypersensitive necrosis), but also with adjacent host cells, perhaps 4–5 cell layers away (Fig. 3). This requirement exists because of the inadequacy of active defence occurring only in the initially contacted host cell. Induced defence involves a time lag of a few hours after initial recognition before pathogen growth is inhibited (*ca* 7 h in the soybean–*P. megasperma* f.sp. *glycinea* system (Yoshikawa *et al.*, 1978a, b)). If only the initially contacted plant cells responded, the fungus would have grown well beyond them before they could accumulate phytoalexin concentrations sufficient to stop its growth. With a systemic second messenger, phytoalexin production could occur in host cells well in advance of the invading pathogen, as has been experimentally observed to occur in resistant plants (Yoshikawa *et al.*, 1978b).

The occurrence of *de novo* DNA transcription and RNA translation predicted in Fig. 3 is supported by several lines of evidence. Specific inhibitors of these processes block the occurrence of HR in response to inoculation with an incompatible pathogen race which then develops for some time as if it were genetically compatible (Doke *et al.*, 1975; Tani and Yamamoto, 1979; Yoshikawa *et al.*, 1978a, b). Using UV light and photoreversibility, Reilly and Klarman (1980) provided additional direct evidence that elicitation of glyceollin in soybean cells involved DNA transcription. Further, m-RNA biosynthesis was stimulated more in soybean hypocotyls inoculated with an incompatible race of *Phytophthora megasperma* f.sp. *glycinea* than with a compatible race (Yoshikawa *et al.*, 1977), and new proteins were synthesized in a wheat germ system using m-RNA from incompatible but not compatible inoculated hypocotyls (Yoshikawa, 1980). Two of the key enzymes in glyceollin biosynthesis, specific C-prenyl transferases, were also inducibly formed in soybean cotyledons following elicitor treatment (Zähringer *et al.*, 1979; 1981), but were not detected in control cotyledons that were only wounded. This demonstrated that at least two enzymes in the terminal biosynthetic pathway to glyceollin are inducible, as predicted by the elicitor–receptor model (Fig. 3).

A final aspect of the elicitor–receptor scheme is the tenet that newly synthesized phytoalexins are transported from the plant cells in which they are synthesized and accumulate in dead, hypersensitive host cells in the pathogen infection court. This short distance systemicity of phytoalexins is theoretically consistent with their role in disease defence, since it allows host cells surrounding the infection site to export phytoalexins and/or their biosynthetic intermediates into the dead hypersensitive cells and thus to facilitate containment of the pathogen (Keen, 1981, 1982). Although there is now substantial reason to believe that such movement of phytoalexins occurs, its mechanism is unknown. In Fig 3, the possibility of extracellular liberation of phytoalexins via vesicles is depicted, but no direct evidence supports this idea.

According to the elicitor–receptor hypothesis, avirulence genes in fungus pathogens probably code for specific glycosyl transferase enzymes, as Albersheim and Anderson predicted in 1975 (Fig. 4). These enzymes catalyse the addition of unique sugar residues in specific linkages to nascent glycosyl chains that are destined to become the carbohydrate side chains of the surface and extracellular glycoproteins of eucaryotes. The glycosyl transferase hypothesis for avirulence gene function in fungal pathogens is supported by the direct evidence discussed above and is also consistent with the genetic and physiologic constrains listed in Tables II and III. Let us examine further how such a mechanism fits the genetical and physiological observations for gene-for-gene systems and with the available information on animal immune systems.

The surface carbohydrate elicitor–receptor hypothesis is consistent with genetical and physiological observations for gene-for-gene systems involving dominant avirulence genes and dominant resistance genes. It can also explain the function of recessive avirulence genes, but is totally

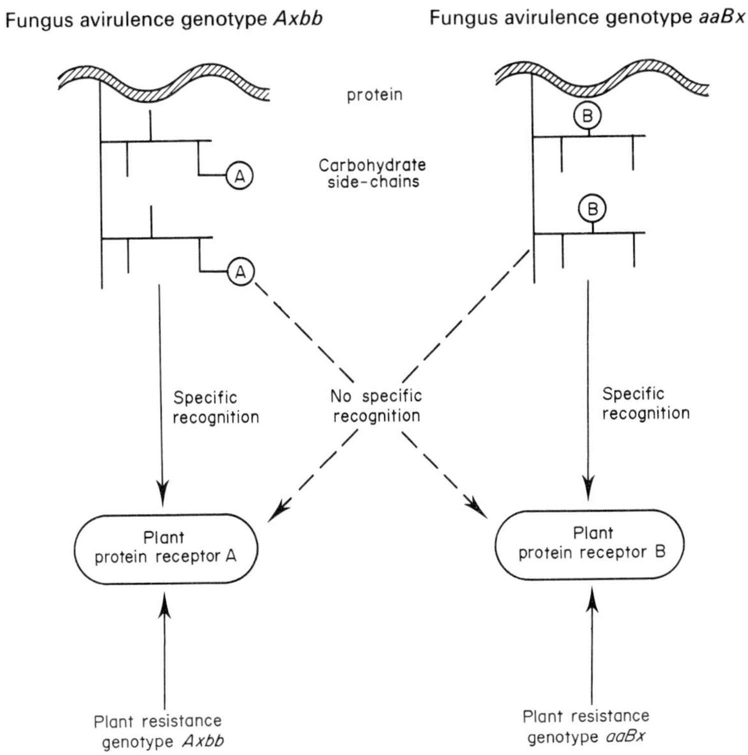

Fig. 4. Schematic representation of the carbohydrate sidechains of fungus surface glycoproteins in two parasite races. Specific sugar residues A or B are added to the carbohydrate chains by specific glycosyl transferase enzymes coded for by avirulence genes A and B, respectively. These sugar residues specifically interact only with the complementary receptors coded for by plant resistance genes A or B, respectively, to initiate host defence.

inconsistent with cases involving recessively inherited resistance genes. Perhaps the mechanism of resistance in these cases is entirely different from that conferred by dominant genes. For example, the resistance of soybean leaves to *Xanthomonas phaseoli* var. *sojensis* is conferred by a single recessive gene, the expression of which does not appear to involve hypersensitivity or phytoalexin accumulation (Fett and Sequeira, 1980; Keen, unpublished).

In the case of dominant resistance genes, the model proposes that the carbohydrates of fungal and possibly bacterial pathogens contain unique sugar residues in specific linkages that are recognized by resistance gene receptors to invoke the HR-phytoalexin system. The pathogen carbohydrates are added to the polysaccharide sidechains by specific glycosyl transferase enzymes, which are in turn coded for by the avirulence genes complementary to host resistance genes (Fig. 4). Recessive pathogen virulence alleles would produce defective transferase proteins incapable of efficiently catalyzing addition of the appropriate sugars to the surface carbohydrates (Fig. 4).

The pathogen elicitors in Fig. 4 are represented as single unique sugar residues that are specifically recognized by complementary plant receptors coded for by resistance alleles. From experience with antigen recognition, however, more complexity is possible, and the receptors may only efficiently recognize two or more sugars in a specific structural configuration. With bacterial and fungal carbohydrate antigens, for example, mammalian immune responses may be directed to single sugar residues or to specific internal oligosaccharide sequences of 2–6 sugar residues.

How can dominant virulence genes or inhibitor genes in the pathogen be rationalized? The model predicts that such genes also code for glycosyl transferases, but these may function in at least two ways: (1) they may add a different sugar(s) to the elicitor active sugar residue(s) (e.g. A or B in Fig. 4) such that the latter is intramolecularly masked and no longer recognized by the plant receptor; (2) they may be responsible for the addition of sugar residues proximal to the carbohydrate residue(s) recognized by the plant receptor; if the proximal sugar residues are not added, then the avirulence gene glycosyl transferase cannot function and the gene is not expressed. The work of Whatley *et al.* (1980) with LPS from smooth *Pseudomonas solanacearum* strains offer precedent for the former mechanism. The latter situation occurs in the biosynthesis of mammalian glycoproteins (Beyer *et al.*, 1979) and in the *mnn*^{3} mutant of *Saccharomyces cerevisiae* (Cohen *et al.*, 1980). It should be noted that example (1) above is a biochemical representation of the suppressor scheme suggested by Bushnell and Rowell (1981).

The elicitor-model can also explain cases of two avirulence genes complementary to one host resistance gene (point 15, Table II). In these cases, situation (2) in the preceding paragraph may occur or, alternatively, two avirulence gene transferases may be required to add two different sugars, both of which must be present for efficient recognition by the resistance gene receptor. Since avirulence gene products are not the recognition elements *per se*, the model predicts that allelism of virulence loci should not commonly occur, as has already been observed (point 14, Table II). The model will also accommodate instances where a single resistance gene confers resistance to incompatible races of related but different pathogen species (point 9, Table II). The organisms in these cases must contain avirulence genes that code for functionally identical glycosyl transferase enzymes.

The elicitor–receptor model predicts a functional basis for stabilizing selection in plant–pathogen systems (point 10, Table II). Insight is supplied by the work of Ballou (1974, 1976) with yeast surface mannan glycoproteins. Several mutant strains of *Saccharomyces cerevisiae* lacking certain distinct carbohydrate residues or linkages in the outer chain polysaccharides have been isolated, but viable mutants lacking all of the outer chain carbohydrates have not been obtained (Delpech *et al.*, 1977; Ballou, 1974). These observations establish that the carbohydrate side chains are important for viability of the fungus and the sidechain structure of the wild-type organism has been selected through evolution rather than one of the reduced mutant types. Assuming that plant pathogens operate in a similar manner, then mutants lacking portions of the carbohydrate outer chains of the surface glycoproteins (viz. virulent races, according to the elicitor–receptor scheme) would be expected to be viable, just as the yeast mutants are. However, these strains might be at a selective disadvantage relative to the wild-type in the absence of selection pressure against an avirulence gene in the wild-type by a host resistance gene. This should also be an environmentally affected process; that is, the loss of fitness associated with the virulence allele may differ depending on environment. Accordingly, in the appropriate environment, weak host resistance genes complement parasite avirulence genes (glycosyl transferases) whose loss in virulent races would produce only a small negative effect on parasite fitness. On the other hand, host resistance genes that are strong would complement glycosyl transferase enzymes in the parasite whose loss would lead to deficient glycosyl sidechains in the surface glycoconjugates giving a more pronounced deleterious effect on parasite fitness.

The model predicts that host resistance genes are structural and that the gene products, probably proteins or glycoproteins, are the recognition elements *per se* (point 6, Table II), distinct from the case for avirulence gene products. This aspect of the model is consistent with most of the points in Table II. For instance, gene doseage effects and gene additivity (point 2, Table II) are accommodated if the quantity of surface receptor molecules resulting from each independent allele is not saturating and does not give a fully resistant response alone. The temperature sensitivity of some resistance genes (point 8, Table II) can be seen to result if the receptor proteins (or glycoproteins) are themselves temperature sensitive, such that they lose conformation and become non-functional at the restrictive temperature.

The model allows the function of both fast and slow resistance genes (point 2, Table II). The most plausible mechanism is for the various resistance gene receptors to have different affinity constants with their respective complementary pathogen elicitor moieties: fast genes would have high affinities and slow genes lower affinity constants. High affinity constants would give more rapid and greater production of the second messenger (Fig. 3), thereby causing a more rapid and more massive invocation of active defence. In addition, factors such as the relative number and geometric availability of elicitor-active moieties in the pathogen and the relative number of receptor molecules in host cells might contribute to the speed and magnitude of the resistant response.

The retention of some function by resistance genes confronted by virulent parasite races (point 11, Table II) is also consistent with the model. The number of elicitor active sugar residues is high in the avirulent race because it possesses a fully active glycosyl transferase; however, the defective glycosyl transferase in a mutant, virulent race may not have entirely lost function. Instead it may function at low efficiency, such that a few elicitor-active carbohydrates are added to the fungus surface glycoconjugates and therefore lead to a low-level defence reaction in the host. A series of allelic virulence genes of this type would be predicted to give different (but low) levels of defence invocation because unique mutations (i.e. different alterations in the primary structure of the relevant glycosyl transferase) should lead to different but substantial losses in catalytic efficiency. Martin and Ellingboe (1976) may have observed this with the three virulent mildew isolates discussed in section IV, but their virulence genes have not been subjected to genetical analysis.

The function of certain mutant avirulence genes would be predicted to be temperature-sensitive, an interpretation consistent with the previously discussed results of Gabriel *et al.* (1979) and Holliday *et al.* (1981).

The model allows the occurrence of mesothetic reactions (point 12, Table II) if it is assumed that the receptor proteins resulting from these genes are not evenly distributed on plant cells. This speculation has credence in view of the widespread phenomenon of capping of surface glycoproteins in animal cells, so that they become concentrated at certain discreet locations on the plasma membrane. Of more interest, production of glycoproteins is frequently developmentally regulated (e.g. Burridge and Jordan, 1979; Roberts *et al.*, 1979; Terhorst *et al.*, 1980): that is, they are produced only during certain portions of the life-cycle or only by certain cell types. A similar phenomenon would explain the mixed reaction types typical of mesothetic reactions if plant cells containing these resistance genes produced quantities of cell surface receptors ranging from high amounts to near zero.

An important consequence of the elicitor–receptor model is that it allows for the phenomenon of allelic resistance genes (point 7, Table II). The model predicts that transcripts of allelic resistance genes produce polypeptides with one or a few amino acid substitutions, and that these possess altered conformations and therefore unique recognition specificities. None of the predicted resistance gene polypeptides has been isolated, but this mechanism operates with the heavy polypeptide chains of mammalian MHC antigens produced by the various mouse H-2 and human HLA alleles (Feinstein, 1979; Orr *et al.*, 1979a, b; Tragardh *et al.*, 1980; Uehara *et al.*, 1980). The receptor–elicitor model further predicts that the receptor products of closely linked plant resistance genes should show conservation of amino acid sequences. This possibility is supported by recent evidence obtained with products of closely linked MHC genes (e.g. Coligan *et al.*, 1980; Tragardh, *et al.*, 1980; Uehara *et al.*, 1980).

VIII. ADDITIONAL PREDICTIONS AND FURTHER TESTS OF THE MODEL

Despite its consistency with the available genetical and biochemical evidence, additional experimentation will be required to establish the validity and generality of the elicitor–receptor model in gene-for-gene systems. Fortunately, the model makes several predictions which should be amenable to further testing. Some of the questions for future consideration are listed in Table VI. Foremost is the need to search for the hypothesized specific receptors in plant tissues, especially plasma membrane fractions. To effectively do this using affinity techniques, purified and characterized pathogen-specific elicitors must be available. Based on the progress to date, it is reasonable to suppose that this will be possible in the near future. Perhaps the pathogen elicitors will also specifically cause detectable responses with living host cells similar to those already demonstrated with non-specific elicitors, including agglutination and killing of protoplasts (Doke and Tomiyama, 1980; Peters *et al.*, 1978). Gabriel and Ellingboe (1978) observed unique proteins on two dimensional gel electrophoretograms from a resistant genotype of wheat that were not present in the near-isogenic susceptible line. These may have been the hypothesized receptor proteins, but it is difficult to critically test this possibility since elicitors have not been isolated from *Erysiphe graminis* f.sp. *tritici*.

The resistance gene receptors may turn out to be similar to the seed lectins that are widely distributed in higher plants (Callow, 1977; Liener, 1976; Sequeira, 1978). Although evidence for the presence of

Table VI. Questions arising from the elicitor–receptor model shown in Figs 3 and 4.

(1) What is the nature of the hypothesized resistance gene-coded receptors?
(2) Are surface carbohydrate elicitors important in specific interactions of other pathogens (such as bacteria and nematodes) and their hosts?
(3) Are the specific carbohydrate elicitors of gene-for-gene pathogens also important as elicitors of non-host (general) hypersensitive resistance in other plants?
(4) What is the physiological function of the specific defence elicitors in the pathogen?
(5) What are the detailed chemical structures of the specific defence elicitors of fungi and other pathogens?
(6) Is there a gene-for-gene host–parasite system in which a culturable pathogen has a usable recombination system for critically testing the role of structural alterations of surface carbohydrates in race specificity?
(7) What are the hypothesized second messengers?
(8) What is the nature of the molecular events involved in phytoalexin induction, especially how does altered nucleic acid metabolism arise, and how many biosynthetic enzymes are induced?
(9) Can viable plants be obtained which contain mutant genes for the terminal enzymes involved in phytoalexin biosynthesis?

seed lectins in vegetative plant parts is negative in many reports, several papers indicate the presence of glycoproteins associated with plant plasma membranes that appear closely related to seed lectins (Bowles *et al.*, 1979; Gade *et al.*, 1981; Travis and Berkowitz, 1980). These may include the postulated resistance gene receptors as well as perhaps receptors for other effector molecules such as plant growth hormones.

The elicitors from phytopathogenic fungi thus far appear to be carbohydrates. If this is general, we might expect elicitor functions to be associated with bacterial surface carbohydrates and cuticular glycoproteins or polysaccharides of nematodes. Viruses continue to pose a mystery as to how they trigger hypersensitive defence reactions. Despite the high mutability of viruses, it appears that gene-for-gene complementarity is rare in virus–plant systems (see Day, 1974). This may indicate that the specific recognitional strategy of plants for viruses has little in common with that for microbial pathogens.

The surface carbohydrates of pathogens which interact specifically with resistance receptors of their host plant may also function as non-specific elicitors of the HR in certain non-hosts. In addition, the surface carbohydrates of pathogens closely related to those in gene-for-gene systems may function as non-specific defence elicitors in non-host plants. For instance, in addition to blocking resistance of soybeans to incompatible *P. megasperma* f.sp. *glycinea* races, the translation inhibitor blasticidin S also blocked phytoalexin production in soybean hypocotyls inoculated with five non-pathogenic *Phytophthora* spp., and these grew as well as the pathogen (Yoshikawa *et al.*, 1978a). This suggests that surface carbohydrates of the related fungi non-specifically elicit phytoalexin production in soybean, thus constituting general resistance. It would therefore be interesting to isolate the elicitors from various *Phytophthora* spp. and to test their effects in soybean.

What is the physiological function of the surface carbohydrates of plant pathogens? Despite the considerable amount of structural variability that may occur in the outer chain carbohydrates of fungus surface glycoproteins and Gram negative bacterial LPS, mutants lacking all of the carbohydrates have not been obtained (Ballou, 1974). Thus, the surface carbohydrates must be essential for viability. In more recent work, Ballou *et al.* (1980) reported that a glycosyl transferase deficient yeast mutant with reduced outer chain carbohydrate structures exhibited slower growth, disturbed physiology and tended to lyse abnormally. The precise role of the surface carbohydrates remains obscure, but they have been shown to be important in conferring mating-type reactions (Crandall *et al.*, 1977; Kawanabe *et al.*, 1978), in spore survival (Nicholson and Maraes, 1980) and in maintenance of wild-type morphology (Delpech *et al.*, 1977). These findings suggest that various structural changes in the outer chain carbohydrates may have different effects on the biological fitness of the organism. This may reflect the situation seen with fungal pathogens in stabilizing selection: some virulent races may suffer greater losses of fitness in a certain environment than races resulting from mutations at other avirulence gene loci.

Additional degrees of complexity doubtless remain to be encountered. For instance, it is possible that the surface elicitors of some plant pathogens may be developmentally controlled, as noted above for certain glycoproteins of micro-organisms and animal cells. Resistance to some pathogens appears to be expressed only when a certain developmental state is attained (e.g. formation of first haustoria by rust fungi). Perhaps the elicitors recognized by the host in these cases are not present on infection hyphae but only on haustorial mother cells. It is also possible that the surface elicitors formed by pathogens in their host tissue may differ qualitatively or quantitatively from those formed in culture. It is noteworthy in this that developmental effects on the structure of surface carbohydrates from certain rhizobia are thought to affect bacterial nodulation of legume roots (Bauer, 1981; Hrabak *et al.*, 1981).

Considering the recent indications for the importance of pathogen surface carbohydrates in gene-for-gene systems, there should be expanded efforts to elucidate their structures. Sequencing of complex carbohydrates has lagged behind that of proteins and nucleic acids, but recent advances should facilitate structural investigation of pathogen surface and extracellular carbohydrates (e.g. Valent *et al.*, 1980).

A remaining probem is the lack of usable recombinational systems in the phytopathogenic fungi from which surface carbohydrate elicitors have been studied. To critically establish the precise function of avirulence genes, the capability to perform genetic crosses of virulent and avirulent races and to obtain induced mutants would be most useful, but is thus far an unattained goal with pathogens that are amenable to biochemical studies.

We know nothing of the nature of the second messengers hypothesized to interact between the plasma membrane receptors and the nucleus of the contacted and adjacent host cells (Fig. 4). Whether these turn out to be second messengers similar to those from other biological systems or are themselves carbohydrates such as the endogenous plant elicitors described by Albersheim *et al.* (1980) is unresolved.

Despite several advances in the last few years, we need more research to test critically the role of *de novo* transcription and translation in resistant reactions predicted by the elicitor–receptor model: what is the precise nature of regulation of transcription; are the *de novo* messenger RNAs processed and are they super-abundant; and how many phytoalexin biosynthetic enzymes are inducibly formed following challenge of plant tissue with an incompatible pathogen race?

If the elicitor–receptor model is correct, it should be possible to obtain plants carrying natural or induced mutations of genes coding for terminal enzymes in phytoalexin biosynthetic pathways (Ellingboe, 1982; Keen, 1982). Such mutants would also be useful in independently assessing the role of phytoalexin production in disease resistance. However, phytoalexin-less mutant plants have not been observed, possibly because any spontaneous mutants are destroyed by pathogens and therefore are not viable in nature. Their isolation may require the use of cell culture techniques and the maintainance of mutant plants in gnotobiotic environments.

It is perhaps appropriate to end this paper with the preceding questions since there are more uncertainties than answers in our understanding of recognition in gene-for-gene systems. Our greatest deficiency is the absence of direct information on the nature of avirulence and resistance gene products. However, the arrival of recombinant DNA techniques portends discoveries of considerable biologic significance in this practically important area of research.

ACKNOWLEDGEMENTS

The author's research is supported by the National Science Foundation. M. Holliday, M. Yoshikawa and A. I. Zaki made useful comments on the manuscript. L. Browder, G. Statler, A. Ellingboe, A. Roelfs, G. Green, R. McIntosh, G. Mayo, P. Day and M. Wolfe supplied valuable genetic information.

REFERENCES

Albersheim, P. and Anderson-Prouty, A. J. (1975). *Annu. Rev. Plt. Physiol.* **26,** 31–52.

Albersheim, P., Darvill, A. G., McNeil, M., Valent, B. S., Hahn, M. G., Lyon, G., Sharp, J. K., Desjardins, A. E., Spellman, M. W., Ross, L. M., Robertsen, B. K., Aman, P. and Franzen, L.-E. (1980). *Proc. X Int. Symp. Carbohydrate Chem.*, Sydney, Australia, July, 1980.

Anderson, A. J. (1980). *Can. J. Microbiol.* **26,** 1473–1479.

Ayers, A. R., Ayers, S. B. and Goodman, R. N. (1979). *Appl. Environ. Microbiol.* **38,** 659–666.

Ayers, A. R., Ebel, J., Valent, B. and Albersheim, P. (1976). *Plant Physiol.* **57,** 760–765.

Bailey, J. A., Rowell, P. M. and Arnold, G. N. (1980). *Physiol. Plant Pathol.* **17,** 329–339.

Baker, E. P. (1966). *Euphytica* **15,** 313–318.

Ballou, C. E. (1974). *Adv. Enzymol.* **40,** 239–270.

Ballou, C. (1976). *Adv. Microbiol. Physiol.* **14,** 93–158.

Ballou, L., Cohen, R. E. and Ballou, C. E. (1980). *J. Biol. Chem.* **255,** 5986–5991.

Bateman, D. F. and Beer, S. V. (1965). *Phytopathology* **55,** 204–211.

Barkai-Golan, R., Mirelman, D. and Sharon, N. (1978). *Arch. Microbiol.* **116,** 119–124.

Beyer, T. A., Rearick, J. I., Paulson, J. C., Prieels, J.-P., Sadler, J. E. and Hill, R. L. (1979). *J. Biol. Chem.* **254,** 12531–12541.

Bowles, D. J., Lis, H. and Sharon, N. (1979). *Planta* **145,** 193–198.

Bruegger, B. B. and Keen, N. T. (1979). *Physiol. Plant Pathol.* **15,** 43–51.

Burns, W. H. and Allison, A. C. (1975). *In* "The antigens", Vol. III. (M. Sela., ed.), pp. 479–574. Academic Press, New York and London.

Burridge, K. and Jordan, L. (1979). *Exp. Cell Res.* **124,** 31–38.

Bushnell, W. R. and Rowell, J. B. (1981). *Phytopathology* **71,** 1012–1014.

Callow, J. A. (1977). *Adv. Bot. Res.* **4,** 1–49.

Calub, A. G., Dunn, G. M. and Routley, D. G. (1973). *Crop Sci.* **13,** 561–563.

Chakravarti, B. P. (1966). *Phytopathology* **56,** 223–229.

Christ, B. J. and Groth, J. V. (1979). *Phytopathology* **69,** 1024. (abstract).

Coffey, M. D. (1976). *Can. J. Bot.* **54,** 1443–1457.

Cohen, R. E., Ballou, L. and Ballou, C. E. (1980). *J. Biol. Chem.* **255,** 7700–7707.

Coligan, J. E., Kindt, T. J., Nairn, R., Nathenson, S. G., Sachs, D. H. and Hansen, T. H. (1980). *Proc. Nat. Acad. Sci. USA* **77,** 1134–1138.

Crandall, M., Egel, R. and MacKay, V. L. (1977). *Adv. Microbiol. Physiol.* **15,** 307–398.

Crill, P. (1977). *Annu. Rev. Phytopathol.* **15,** 185–202.

Crosthwaite, L. M. and Patil, S. S. (1974). *Phytopathol. Z.* **91,** 80–90.

Cruickshank, I. A. M. (1980). *In* "Plant Disease", Vol. V. (J. G. Horsfall and E. B. Cowling, eds) pp. 247–267. Academic Press, New York and London.

Daly, J. M. (1972). *Phytopathology* **62,** 392–400.

Day, P. R. (1974). "Genetics of host–parasite interaction", W. H. Freeman, San Francisco. 238 p.

Dazzo, F. B., Yanke, W. E. and Brill, W. J. (1978). *Biochim. Biophys. Acta* **539,** 276–286.

DeFago, G. (1982). *In* "Active defense mechanisms in plants" (R. K. S. Wood, ed.), Plenum Press, London.

Delpech, I., Bizeau, C., Bonaly, R. and Galzy, P. (1977). *Ann. Microbiol* **128B,** 439–450.

deWit, P. J. G. M. and Flach, W. (1979). *Physiol. Plant Pathol.* **15,** 257–267.

deWit, P. J. G. M. and Roseboom, P. H. M. (1980). *Physiol. Plant Pathol.* **16,** 391–408.

Doke, N., Nakae, Y. and Tomiyama, K. (1975). *Phytopathol. Z.* **87,** 337–344.

Doke, N. and Tomiyama, K. (1978). *Physiol. Plant Pathol.* **12,** 133–139.

Doke, N. and Tomiyama, K. (1980). *Physiol. Plant Pathol.* **16,** 169–176.

Dowley, L. J., Routley, D. G. and Peirce, L. C. (1975). *Phytopathology* **65,** 1422–1424.

Dunn, G. M. and Namm, T. (1970). *Crop Sci.* **10,** 352–354.

Dyck, P. L. and Samborski, D. J. (1970). *Can. J. Genet. Cytol.* **12,** 689–694.

Dyck, P. L. and Samborski, D. J. (1974). *Can. J. Genet. Cytol.* **16,** 323–332.

Ebba, T. and Person, C. (1975). *Can. J. Genet. Cytol.* **17,** 631–636.

Ellingboe, A. H. (1976). *In* "Physiological Plant Pathology", (R. Heitefuss and P. H. Williams, eds), pp. 761–778. Springer-Verlag, Berlin.

Ellingboe, A. H. (1978). *In* "The Powdery Mildews", (J. M. Spencer, ed.) pp. 159–181. Academic Press, London and New York.

Ellingboe, A. H. (1980). *In* "Biology of Plant Productivity", pp. 203–229 Academic Press, New York and London (in press).

Ellingboe, A. H. (1981). *Proc. IX Cong. Plt. Protection Vol I.* (T. Kommendahl, ed.) 194–197. Burgess, Minneapolis.

Ellingboe, A. H. (1982). *In* "Active Defense Mechanisms in Plants", (R. K. S. Wood, ed.) Plenum Press.

Elnagy, M. A. and Heitefuss, R. (1976). *Physiol. Plant Pathol.* **8,** 269–277.

Fett, W. F. and Sequeira, L. (1980). *Plant Physiol.* **66,** 853–858.

Feinstein, A. (1979). *Nature* **282,** 230.

Flor, H. H. (1971). *Annu. Rev. Phytopathol.* **9,** 275–296.

Friend, J. (1979). *Recent Adv. Phytochem.* **12,** 557–588.
Gabriel, D. W. and Ellingboe, A. H. (1978). *Phytopathol. News* **12,** 139.
Gabriel, D. W., Ellingboe, A. H. and Rossman, E. C. (1979). *Can J. Bot.* **57,** 2639–2643.
Gade, W., Jack, U. A., Dahl, J. B., Schmidt, E. C. and Wold, F. (1981). *J. Biol, Chem.* **256,** 12905–12910.
Gallun, R. L. and Khush, G. S. (1980). *In* "Breeding plants resistant to insects", (F. G. Maxwell and P. R. Jennings, eds), pp. 63–85. Wiley-Interscience, New York.
Gnanamanickam, S. S. and Patil, S. S. (1977). *Physiol. Plant Pathol.* **10,** 159–168.
Goodman, R. N., Huang, P-Y. and White, J. A. (1976). *Phytopathology* **66,** 754–764.
Graham, T. L., Sequeira, L. and Huang, T-S. R. (1977). *Appl. Environ. Microbiol.* **34,** 424–432.
Green, G. J. (1970). *Can. J. Bot.* **48,** 969–976.
Green. G. J. and McKenzie, R. I. H. (1967). *Can. J. Genet. Cytol.* **9,** 785–793.
Hahn, M. G., Darvill, A. G. and Albershein, P. (1981). *Plant Physiol.* **68,** 1161–1169.
Harder, D. E., Rohringer, R., Samborski, D. J., Rimmer, S. R., Kim, W. K. and Chong, J. (1979). *Can. J. Bot.* **57,** 2617–2625.
Hargreaves, J. A. and Bailey, J. A. (1978). *Physiol. Plant Pathol.* **13,** 89–100.
Haywood, M. J. and Ellingboe, A. H. (1979). *Phytopathology* **69,** 48–53.
Heath, M. C. (1974). *Physiol. Plant Pathol.* **4,** 403–414.
Heath, M. C. (1980). *Annu. Rev. Phytopathol.* **18,** 211–236.
Helgeson, J. P., Budde, A. D. and Haberlach, G. T. (1978). *Plant Physiol.* **61,** Suppl. p. 58 (Abstract).
Himmelhoch, S., Kisiel, M. J. and Zuckerman, B. M. (1977). *Exper. Parasitol.* **41,** 118–123.
Hirata, F. and Axelrod, J. (1980). *Science* **209,** 1082–1090.
Holliday, M. J., Long, M. and Keen, N. T. (1981). *Physiol. Plant Pathol.* **19,** 209–216.
Hooker, A. L. (1967). *Annu. Rev. Phytopathol.* **5,** 163–182.
Hrabak, E. M., Urbano, U. R. and Dazzo, F. B. (1981). *J. Bacteriol.* **148,** 697–711.
Hümme, B., Hoppe, H. H. and Heitefuss, R. (1978). *Phytopathol. Z.* **92,** 281–284.
Hyde, P. M. and Colhoun, J. (1975). *Phytopath. Z.* **82,** 185–206.
Janeway, C. A., Jr. (1980). *Cont. Top. Immunobiol.* **10,** 171–203.
Jann, K. and Westphal, O. (1975). *In* "The Antigens", (M. Sela, Ed.). Vol. III, pp. 1–125. Academic Press, New York and London.
Johnson, G., Maag, D. D., Johnson, D. K. and Thomas, R. D. (1976). *Physiol. Plant Pathol.* **8,** 225–230.
Jones, F. G. W. and Parrott, D. M. (1965). *Ann. Appl. Biol.* **56,** 27–36.
Jones, D. R. and Deverall, B. J. (1977). *Physiol. Plant Pathol.* **10,** 275–284.

Kaplan, D. T., Keen, N. T. and Thomason, I. J. (1980). *Physiol. Plant Pathol.* **16,** 309–318.

Kawanabe, Y., Hagiya, M., Yoshida, K. and Yanagishima, N. (1978). *Plant Cell Physiol.* **19,** 1207–1216.

Keen, N. T. (1971). *Physiol. Plant Pathol.* **1,** 265–275.

Keen, N. T. (1975). *Science* **187,** 74–75.

Keen, N. T. (1978a). *Phytopathology* **68,** 1237–1239.

Keen, N. T. (1978b). *Phytopathol. News* **12,** 221 (Abstr.).

Keen, N. T. (1981). *In* "Plant Disease Control: Resistance and susceptibility", (R. Staples and G. Toeniessen, eds) pp. 155–177. Wiley Interscience, New York.

Keen, N. T. (1982). *In* "Active defense mechanisms in plants", (R. K. S. Wood, ed.) Plenum Press.

Keen, N. T. and Bruegger, B. B. (1977). *Amer. Chem. Soc. Symp. Ser.* **62,** 1–26.

Keen, N. T. and Holliday, M. J. (1982). *In* "Recognition of Bacterial Pathogens by Plants", Vol. II (M. Mount and G. Lacy, eds), Academic Press, New York and London (in press).

Keen, N. T., Ersek, T., Long, M., Bruegger, B. and Holliday, M. (1981). *Physiol. Plant Pathol.* **18,** 325–337.

Keen, N. T. and Kennedy, B. W. (1974). *Physiol. Plant Pathol.* **4,** 173–185.

Keen, N. T. and Legrand, M. (1980). *Physiol. Plant Pathol.* **17,** 175–192.

Keen, N. T. and Littlefield, L. J. (1979). *Physiol. Plant Pathol.* **14,** 265–280.

Keen, N. T., Partridge, J. E. and Zaki, A. I. (1972). *Phytopathology* **62,** 768 (abstract).

Kenny, G. E. (1975). *In* "The Antigens", Vol. III (M. Sela, ed.) pp. 449–478. Academic Press, New York and London.

Klein, J. (1979). *Science* **203,** 516–521.

Klement, Z. (1977). *Acta Phytopath. Acad. Sci. Hung.* **12,** 257–261.

Kojima, M. and Uritani, I. (1974). *Plant Cell Physiol.* **15,** 733–737.

Kritzman, G., Chet, I. and Henis, Y. (1977). *Exp. Mycol.* **1,** 280–285.

Kuehl, F. A., Jr. and Egan, R. W. (1980). *Science* **210,** 978–984.

Law, C. N. and Johnson, R. (1967). *Can. J. Genet. Cytol.* **9,** 805–822.

Lawrence, G. J., Mayo, G. M. E. and Shepherd, K. W. (1981). *Phytopathology* **71,** 12–19.

Lazarovits, G. and Higgins, V. J. (1976). *Can. J. Bot.* **54,** 224–234.

Lehle, L. (1980). *Eur. J. Biochem.* **109,** 589–601.

Leonard, K. J. and Czochor, R. J. (1980). *Annu. Rev. Phytopathol.* **18,** 237–258.

Liener, I. E. (1976). *Ann. Rev. Plant Physiol.* **27,** 291–319.

Lim, S. M., Hooker, A. L. and Paxton, J. D. (1970). *Phytopathology* **60,** 1071–1075.

Littlefield, L. J. (1969). *Phytopathology* **59,** 1323–1328.

Littlefield, L. J. (1973). *Physiol. Plant Pathol.* **3,** 241–247.

Littlefield, L. J. and Aronson, S. J. (1969). *Can. J. Bot.* **47,** 1713–1717.

Lyon, R. M. and Wood, R. K. S. (1975). *Physiol. Plant Pathol.* **6,** 117–124.

Mansfield, J. W., Hargreaves, J. A. and Boyle, F. C. (1974). *Nature* **252,** 316–317.

Mares, D. J. (1979). *Physiol. Plant Pathol.* **15,** 289–296.
Martens, J. W., Burnett, P. A. and Wright, G. M. (1977). *Phytopathology* **67,** 1519–1521.
Martens, J. W., McKenzie, R. I. H. and Green, G. J. (1970). *Can J. Bot.* **48,** 969–975.
Martin, T. J. and Ellingboe, A. H. (1976) *Phytopathology* **66,** 1435–1438.
Mayama, S., Daly, J. M., Rehfeld, D. W. and Daly, C. R. (1975). *Physiol. Plant Pathol.* **7,** 35–47.
May Mayama, S., Tani, T., Matsuura, Y., Ueno, T. and Fukami, H. (1981). *Physiol. Plant Pathol.* **19,** 217–226.
Mayo, G. M. E. and Shepherd, K. W. (1980). *Heredity* **44,** 211–227.
Mazzucchi, U. and Pupillo, P. (1976) *Physiol. Plant Pathol.* **9,** 101–112.
McIntosh, R. A. (1976). *J. Austr. Inst. Agric. Sci.* pp. 203–216.
McIntosh, R. A. (1977). *In* "Induced mutations against plant diseases", pp. 551–565. *Int. Atomic Energy Agency*, Vienna, 1977.
Nakajima, T., Sasaki, H., Sato, M., Tamari, K. and Matsuda, K. (1977). *J. Biochem.* **82,** 1657–1662.
Nass, H. A., Pedersen, W. L., MacKenzie, D. R. and Nelson, R. R. (1981). *Phytopathology* **71,** 1315–1318.
Nelson, R. R. (1979). *In* "Host-parasite interfaces", (B. B. Nickel, ed.) pp. 17–25. Academic Press, New York and London.
Nelson, R. R., MacKenzie, D. R. and Scheifele, G. L. (1970). *Phytopathology* **60,** 1250–1254.
Nichols, E. J., Beckman, J. M. and Hadwiger, L. A. (1980). *Plant Physiol.* **66,** 199–204.
Nicholson, R. L. and Moraes, W. B. C. (1980). *Phytopathology* **70,** 255–261.
Nozue, M., Tomiyama, K. and Doke, N. (1979). *Physiol. Plant Pathol.* **15,** 111–115.
Oku, H., Ouchi, S., Shiraishi, T., Baba, T. and Miyagawa, H. (1975a). *Proc. Jap. Acad.* **51,** 198–201.
Oku, H., Ouchi, S., Shiraishi, T., Komoto, Y. and Oki, K. (1975b). *Ann. Phytopathol. Soc. Japan* **41,** 185–191.
Orr, H. T., Lopez de Castro, J. A., Parham, P., Ploegh, H. L. and Strominger, J. L. (1979a). *Proc. Nat. Acad. Sci. USA* **76,** 4395–4399.
Orr, H. T., Lancet, D., Robb, R. J., Lopez de Castro, J. A. and Strominger, J. L. (1979b). *Nature* **282,** 266–270.
Ouchi, S., Hibino, C. and Oku, H. (1976). *Physiol. Plant Pathol.* **9,** 25–32.
Ouchi, S., Oku, H., Hibino, C. and Akiyama, I. (1974). *Phytopathol.* **79,** 142–154.
Paddock, W. C. (1953). *New York (Cornell) Ag. Exp. Sta. Mem.* **315,** p. 63.
Parlevliet, J. E. (1976). *Euphytica* **25,** 249–254.
Parlevliet, J. E. (1978). *Euphytica* **27,** 369–379.
Parlevliet, J. E. (1979). *Annu. Rev. Phytopathol.* **17,** 203–222.
Partridge, J. E. and Keen, N. T. (1976). *Phytopathology* **66,** 426–429.
Paxton, J. D. (1980). *Plant Disease* **64,** 734.
Paxton, J. D. and Chamberlain, D. W. (1967). *Phytopathology* **57,** 352–353.

Person, C., Groth, J. V. and Mylyk, O. M. (1976). *Annu. Rev. Phytopathol.* **14,** 177–188.

Peters, B. M., Cribbs, D. H. and Stelzig, D. A. (1978). *Science* **201,** 364–365.

Philipp, M., Parkhouse, R. M. E. and Ogilvie, B. M. (1980). *Nature* **287,** 538–540.

Prusky, D., Dinoor, D and Jacoby, B. (1980). *Physiol. Plant Pathol.* **17,** 33–40.

Rahe, J., Kuc, J., Chuang, C.-M. and Williams, E. B. (1969). *Neth. J. Plant Pathol.* **75,** 58–71.

Rebois, R. V., Epps, J. M. and Hartwig, E. E. (1970). *Phytopathology* **60,** 695–700.

Reilly, J. J. and Klarman, W. L. (1980). *Environ. Expt. Bot.* **20,** 131–134.

Riley, R. (1973). *Ann. Appl. Biol.* **75,** 128–132.

Roberts, I. N., Stead, A. D., Ockendan, D. J. and Dickinson, H. G. (1979). *Planta* **146,** 179–183.

Robinson, R. A. (1969). *Rev. Appl. Mycol.* **48,** 593–606.

Robinson, R. A. (1980). *In* "Breeding plants resistant to insects", (F. G. Maxwell and P. R. Jennings eds), pp. 157–181. Wiley Interscience, New York.

Rodrigues, C. J., Jr., Madeiros, E. F. and Lewis, B. G. (1975). *Physiol. Plant Pathol.* **6,** 35–41.

Royle, D. J. (1976). *In* "Biochemical aspects of plant-parasite relationships", (J. Friend and D. R. Threlfall eds), pp. 161–193. Academic Press, London and New York.

Rohringer, R., Kim, W. K. and Sambroski, D. J. (1979). *Can. J. Bot.* **57,** 324–331.

Russell, G. E. (1978). Plant Breeding for pest and disease resistance. Butterworths, London 485 p.

Samborski, D. J. (1963). *Can. J. Bot.* **41,** 475–479.

Samborski, D. J. and Dyck, P. L. (1968). *Can. J. Genet. Cytol.* **10,** 24–32.

Samborski, D. J. and Dyck, P. L. (1976). *Can. J. Bot.* **54,** 1666–1671.

Sato, N., Kitazawa, K. and Tomiyama, K. (1971). *Physiol. Plant Pathol.* **1,** 289–295.

Saxena, K. M. S. and Hooker, A. L. (1968). *Proc. Nat. Acad. Sci. U.S.A.* **61,** 1300–1305.

Scheffer, R. P. (1976). *In* "Physiological Plant Pathology", (R. Heitefuss and P. H. Williams, eds), pp. 247–269. Springer-Verlag, Berlin.

Schmidt, E. L. (1979). *Ann. Rev. Microbiol.* **33,** 355–376.

Sequeira, L. (1978). *Ann. Rev. Phytopathol.* **16,** 453–481.

Sequeira, L. (1979). *In* "Host–parasite interfaces", (B. B. Nickol, ed.), pp. 71–84. Academic Press, New York and London.

Sequeira, L. and Graham, T. L. (1977). *Physiol. Plant Pathol.* **11,** 43–54.

Shepherd, K. W. and Mayo, G. M. E. (1972). *Science* **175,** 375–380.

Sidhu, G. S. (1975). *Sci. Prog. Oxford* **62,** 467–485.

Sitton, D. and West, C. A. (1975). *Phytochemistry* **14,** 1921–1925.

Slesinski, R. S. and Ellingboe, A. H. (1969). *Phytopathology* **59,** 1833–1837.

Slesinski, R. S. and Ellingboe, A. H. (1971). *Can J. Bot.* **49,** 303–310.

Smith, W. L. and Ballou, C. E. (1974). *Biochemistry* **13,** 355–361.

Snell, G. D., Dausset, J. and Nathenson, S. (1976). "Histocompatibility", Academic Press, London and New York, 401 pp.

Stakman, E. C. (1915). *J. Agr. Res.* **4,** 193–200.

Stall, R. E. and Cook, A. A. (1979). *Physiol. Plant Pathol.* **14,** 77–84.

Statler, G. D. (1977). *Phytopathology* **67,** 906–908.

Statler, G. D. (1979a). *Phytopathology* **69,** 257–259.

Statler, G. D. (1979b). *Phytopathology* **69,** 661–663.

Sunderwirth, S. D. and Roelfs, A. P. (1980). *Phytopathology* **70,** 634–637.

Tani, T. and Yamamoto, H. (1979). *In* "Recognition and specificity in plant host–parasite interactions", (J. M. Daly and I. Uritani, eds), pp. 273–287. University Park Press, Baltimore.

Terhorst, C., von Agthoven, A., Reinherz, E. and Schlossman, S. (1980). *Science* **209,** 520–521.

Tingey, W. M. and Singh, S. R. (1980). *In* "Breeding plants resistant to insects", (F. G. Maxwell and P. R. Jennings, eds), pp. 87–113. Wiley-Interscience, New York.

Tomiyama, K. (1955). *Ann. Phytopathol. Soc. Japan* **19,** 149–154.

Tragardh, L., Rask, L., Wiman, K., Fohlman, J. and Peterson, P. A. (1980). *Proc. Nat. Acad. Sci. U.S.A.* **77,** 1129–1133.

Travis, R. L. and Berkowitz, R. L. (1980). *Plant Physiol.* **65,** 871–879.

Triantaphyllou, A. C. (1975). *J. Nematol.* **7,** 356–364.

Uehara, H., Ewenstein, B. M., Martinko, J. M., Nathenson, S. G., Coligan, J. E. and Kindt, T. J. (1980). *Biochemistry* **19,** 306–315.

Valent, B. S., Darvill, A. G., McNeil, M., Robertsen, B. K. and Albersheim, P. (1980). *Carbohydrate Res.* **79,** 165–192.

Vance, C. P., Kirk, T. K. and Sherwood, R. T. (1980). *Annu. Rev. Phytopathol.* **18,** 259–288.

Van den Heuvel, J. and Glazener, J. A. (1975). *Neth. J. Pl. Path.* **81,** 125–137.

Van der Plank, J. E. (1963). "Plant diseases, epidemics and control". Academic Press, New York and London.

Van der Plank, J. E. (1975). "Principles of plant infection". Academic Press, New York and London. 216 p.

VanEtten, H. D. and Pueppke, S. G. (1976). *In* "Biochemical aspects of plant-parasite relationships", (J. Friend and D. R. Threlfall eds), pp. 239–289. Academic Press, London and New York.

Varns, J. L., Kuc, J. and Williams, E. B. (1971). *Phytopathology* **61,** 174–177.

Wade, M. and Albersheim, P. (1979). *Proc. Nat. Acad. Sci. USA.* **76,** 4433–4437.

Ward, H. M. (1902). *Ann. Bot.* **16,** 233–315.

West, C. A. (1981). *Naturwissenschaften* **68,** 447–457.

West, C. A., Wade, M., McMillan, C., III and Albersheim, P. (1980). *Arch. Biochem. Biophys.* **201,** 25–35.

Whatley, M. H., Hunter, N., Cantrell, M. A., Hendrick, C., Keegstra, K. and Sequeira, L. (1980). *Plant. Physiol.* **65,** 557–559.

Wolfe, M. and Barrett, J. (1981). *Adv. Plant Pathology*, (in preparation).

Wroblewski, H. (1979). *J. Bacteriol.* **140,** 738–741.

Yoshikawa, M. (1980). *Plant Protection* (Tokyo) **34,** 248–256.
Yoshikawa, M., Matama, M. and Masago, H. (1981). *Plant Physiology* **67,** 1032–1035.
Yoshikawa, M., Masago, H. and Keen, N. T. (1977). *Physiol. Plant Pathol.* **10,** 125–138.
Yoshikawa, M., Yamauchi, K. and Masago, H. (1978a). *Plant Physiol.* **61,** 314–317.
Yoshikawa, M., Yamauchi, K. and Masago, H. (1978b). *Physiol. Plant Pathol.* **12,** 73–82.
Zähringer, U., Ebel, J., Mulheirn, L. J., Lyne, R. L. and Grisebach, H. (1979). *FEBS Lett.* **101,** 90–92.
Zähringer, U., Schaller, E. and Grisebach, H. (1981). *Z. Naturforsch.* **36c,** 234–241.
Ziegler, E. and Albersheim, P. (1977). *Plant Physiol.* **59,** 1104–1110.
Zuckerman, B. M., Kahane, I. and Himmelhoch, S. (1979). *Exptl. Parasitol.* **47,** 419–424.

3

THE CHEMISTRY AND BIOLOGY OF PATHOTOXINS EXHIBITING HOST-SELECTIVITY

J. M. Daly and H. W. Knoche

Department of Agricultural Biochemistry, Institute of Agriculture and Natural Resources, University of Nebraska, Lincoln, Nebraska

I. Introduction
  A. Historical
  B. Scientific significance
  C. Economic importance
  D. Definition of terms
II. Production and assay
III. Determination of structure
  A. HV toxin
  B. PC toxin
  C. HC toxin
  D. HS toxin
  E. AK toxin
  F. AM toxin
  G. T toxin
  H. AAl toxin
  I. Other toxins
  J. Summary
IV. Modes of action
  A. Some perspectives
  B. Postulated mechanisms
V. Considerations for future research
  A. On modes of action
  B. On the nature of specificity
  C. On roles in pathogenesis and parasitism
VI. Addendum, September 1981
  References

---

*Published with the approval of the Director as Journal Series Paper No. 6562, Nebraska Agricultural Experiment Station. Some unpublished data presented in this article were obtained through funds from Grant No. 79-00273, CRGO, SEA, US Department of Agriculture and NSF Grant No. PCM 7920685. Literature review completed by October, 1980.

I. INTRODUCTION

A. Historical

IN the early 1940s, a major obstacle in oat cultivation, crown rust disease, seemed to have been overcome with the rapid introduction of new cultivars containing a major gene for resistance derived from the cultivar Victoria. It has been estimated that cultivars with this gene constituted 80% of the US oat acreage by the mid 1940s. Within three years, an epidemic of root and seedling blight caused by *Helminthosporium victoriae* Meehan and Murphy required replacement of them all, an inconceivably rapid and dramatic reversal of the sustained efforts of plant pathogists and geneticists for over a decade. In 1947, Meehan and Murphy in a classic paper attributed the virulence of *H. victoriae* to production of a metabolite which reproduced (a) the wilting and necrosis of natural infections when applied to seedlings, and (b) the patterns of resistance and susceptibility of individual varieties. Further, virulence of the pathogen was correlated with ability to produce the toxic metabolite in culture. The relative ease of production in culture made possible the first clear understanding of the potential role of host-specific toxins in natural disease.

The second well-documented report of a similar phenomenon was that of Hiroe *et al.* in 1958 for *Alternaria kikuchiana* affecting pears. Because of difficulties in scientific communication during this period, the finding was not widely recognized, nor was the very early work of Tanaka (1933) with the same disease. Scheffer and Pringle (1961) first drew widespread attention to a third instance of a toxin involved in specific pathogenicity when they showed that the soil pathogen *Periconia circinata* produced a factor responsible for necrosis on susceptible, but not on resistant, sorghum cultivars.

Since 1961, the list of diseases in which these sorts of fungal metabolites seem important has grown to twelve (Table I), the most recent of which is the improved proof for a toxin of *A. citri* affecting rough lemon, but not mandarin orange (Kohomoto *et al.*, 1979). The same workers also confirmed the existence (Pegg, 1966) of a distinct toxin from the same species, but affecting only mandarin orange. This is the first case of a single species producing more than one host-specific toxin, a fundamental finding for understanding the biology of plant disease.

B. Scientific Significance

A special point to consider in Table I is the report by Lee in 1929 of a

Table I. Demonstrations of host-specific pathotoxins.

Date	Organism and Toxin	Susceptible Species	Discoverer
1947	*Helminthosporium victoriae* (HV toxin)	oats	Meehan and Murphy
1958	*Alternaria kikuchiana* (AK toxin)	pear	Hiroe *et al.* (Tanaka, 1933)[a]
1961	*Periconia circinata* (PC toxin)	sorghum	Scheffer and Pringle
1965	*Helminthosporium carbonum* (HC toxin)	corn	Scheffer and Ullstrup
1966	*Alternaria mali* (AM toxin)	apple	Sawamura
1966	*Alternaria citri*	mandrin orange	Pegg
1970	*Helminthosporium maydis* (T toxin)	corn	Hooker *et al.* (Smedegard-Peterson & Nelson, 1969)[a]
1971	*Helminthosporium sacchari* (HS toxin)	sugar cane	Steiner and Byther (Lee, 1929)[a]
1973	*Phyllosticta maydis* (T toxin)	corn	Comstock *et al.*, Yoder
1976	*Alternaria alternata* f. sp. *lycopersici* (AAl toxin)	tomato	Gilchrist and Grogan
1977	*Pyrenophora teres*[b]	barley	Smedegard-Peterson
1979	*Alternaria citri*	rough lemon	Kohmoto *et al.*
1979	*Alternaria alternata*	strawberry	Maelawa

[a] Earlier reports, neglected through oversight or incomplete evidence, are in parenthesis.
[b] host-selective, not host-specific

factor in cultures of *H. sacchari* causing specific runner symptoms on sugar-cane, and that of Tanaka *et al.* (1933) for the toxin later investigated by Hiroe *et al.* (1958). Whether these reports were not pursued through oversight or through scepticism is not clear. The role of toxins in disease has a long history of contention. The discovery of Meehan and Murphy (1947) faced initial scepticism and, when once generally accepted, was treated as a rare (perhaps the only) example of the chemical basis for specific virulence in disease reaction.

It should be acknowledged that current scientific recognition of the importance of host-specific toxins in disease, and the principles of disease that they may illuminate, are the results of persistent long-term efforts by only a few individuals. H. Wheeler and R. P. Scheffer, for example, separately recognized the potential significance of victorin as a model in studies of specificity in disease. Scheffer and his associates were responsible for the demonstration that victoria blight of oats was not a unique disease. As indicated by Scheffer and Pringle (1967), the demonstration of PC and HC toxins resulted from a deliberate search for such toxins based on the biology of infections caused by *P. circinata* and *H. carbonum.*

It might be argued that a list of only 12 established host-specific toxins is not particularly impressive in view of the large numbers of plant diseases that exist. New discoveries have come somewhat more readily in the last decade (Table I), possibly because of increased acceptability of the concept. The list is probably not more extensive because of the requirement for production in culture to satisfy the protocol currently considered necessary for proofs of the role of pathotoxins in disease. There are undoubtedly many diseases *in vivo* in which host specific toxins play a significant role, but for which appropriate *in vitro* culture systems have not been developed. For many diseases, there also may be no economic justification for extensive research programmes on mechanisms of pathogenesis which may involve host-specific toxins. The *Helminthosporium* diseases of turf grasses are instances where symptoms are suggestive of toxins, but there are few papers dealing with pathogenesis.

The distribution in taxonomic groups of the known examples (Table I) merits comment. Only two pathogenic species outside *Alternaria* and *Helminthosporium* are known to produce a host-specific toxin. This only may reflect the absence of metabolic constraints for toxin production by *Alternaria* and *Helminthospirium* in culture. However, it also may indicate that the ability to cause economic disease in these species is dependent on their innate, and perhaps unique, ability to produce host-specific chemicals. Production of host-specific toxins may be a requirement for pathogenesis and parasitism in these two fungal groups and thus not a mere consequence of culture conditions. A second aspect is that all the known toxins from *Helminthosporia* are effective on cereal hosts. It is also of interest that toxin with the same specificity for Texas male sterile (Tms) corn lines as toxin from *H. maydis*, race T, was discovered independently in *Phyllostica maydis* by Yoder (1973) and Comstock *et al.* (1973). No structural work has been reported from *P. maydis*, but it should be noted that the insecticide methomyl shows the same specificity (Koeppe *et al.* 1978), although its structure bears little resemblance to race T toxin from *H. maydis* (Kono and Daly, 1979). On a molar basis, methomyl is far less active (10^{-2}–10^{-3} M) than native race T toxin (10^{-9}–10^{-10} M).

Host-specific toxins offer a unique scientific opportunity to determine a molecular basis for pathogen virulence and plant disease specificity. Equally important is the potential to cast light on the molecular modes of inheritance for disease reaction (Wheeler, 1969). These aspects are considered in detail in later sections.

C. Economic Importance

Despite the shortness of the list of Table I, the economic importance of host-specific toxins is amply demonstrated by only two examples in the United States: *H. victoriae* and *H. maydis* race T, were responsible for major economic disruptions in oat (1940–1948) and corn (1970–1971) production, disasters in US grain production exceeded only by the ravages of *Puccinia graminis tritici* on wheat in some years. Both diseases appeared with unexpected suddenness. Individual affected fields were nearly devasted in short periods of time. Such epidemiological patterns are expected for disease situations where small amounts of toxicant can act at considerable distances from points of infection. We have frequently observed a necrotic runner extending the entire length (40–50 cm) of a corn leaf from a single lesion of *H. maydis* race T at the basal end. Obviously, the economic damage is disproportionate to the initial load of inoculum. It is difficult to arrive at monetary values for the losses from these two diseases. In addition to yield loss, the negation of the previous research efforts to develop the affected varieties and the cost of their replacements must be calculated. A second outcome of the Southern corn leaf blight epidemic was the reintroduction of labour-intensive hand de-tasselling in hybrid seed corn production and the costly development of new mechanical de-tasselling equipment.

In his discussion of the agronomic background for some of the diseases in Table I, Scheffer (1976) suggests two patterns for their occurrence. Black spot of Japanese pear became a problem with the use of a new pear variety arising by mutation. Similarly, Victoria blight of oats and Southern corn leaf blight developed rather rapidly after the introduction of new genetic factors into host crops. Presumably, the pathogens had existed in nature prior to these genetic changes in the host crop. In contrast, *H. carbonum* became a problem on long established corn varieties, suggesting the acquisition of an ability to produce its host-specific toxin. Obviously, pathogens which produce host-specific toxins have the capability of introducing unexpected instability in crop production.

Despite the damage to economic crops, there potentially are economic gains from research on host-specific toxins. For example, deliberate search for diseases of weeds in which host-specific toxins can be demonstrated might lead to weed control with synthetic or natural compounds having minimal impact on the environment.

D. Definition of Terms

When victorin was the only toxin of its class, the term "host-specific" accurately reflected then current knowledge. The susceptibility of oats upon inoculation with the pathogen clearly followed sensitivity to cultural metabolites. Further, the very high dosages required to show effects on resistant oats could be explained as the result of non-specific contaminants in the impure preparations then available. Most convincingly, all the physiologic and visual symptom of disease could be produced by the toxin. Thus, victorin appeared to be a specific requirement for any diseases at all to develop in susceptible oats (Scheffer and Pringle, 1967). The term "host-specific" seemed very appropriate. Yoder and Scheffer (1969) buttressed these concepts with a direct demonstration that victorin was required at the earliest stage of infection. Wheeler and Luke (1963), however, preferred the term "host-selective" because it allows the possibility that victorin affects resistant cells, but only at high concentrations.

As new instances are described, the earlier concepts must be re-examined. *H. maydis* race T, for example, can cause disease on corn with normal cytoplasm. The toxin does not appear to be required for pathogenicity *per se*, but appears to function in causing additional cellular damage in susceptible, but not resistant, tissue at a distance from the localization of the pathogen. It may not be possible in all cases to equate host-specific toxins with primary disease determinates as once seemed to be valid (Scheffer and Pringle, 1967). The implication that non-specific toxins, such as tentoxin, are secondary determinants of disease may have more validity (Scheffer and Pringle, 1967), but non-specific toxins appear to play the same role in pathogenesis as does T toxin, as an added virulence factor. Therefore, both are primary in an economic sense, but are not absolutely necessary for penetration and initial pathogenesis as appeared to be the case for victorin (Yoder and Scheffer, 1969).

The term "host-selective" favoured by Wheeler and Luke (1963) also does not appear adequate in the light of more recent information. Smedgaard-Peterson (1977) reported a toxin from *Pyrenophora* which affects all barley varieties, but no other genera tested. The material appears to be selective for barley, but is not correlated with ability to cause disease or damage in individual barley cultivars. An analogous situation was reported by Lazarovits and Higgins (1979) and Lazarovits *et al.* (1979).

In all cases, the toxins listed in Table I appear to meet the criterion for a pathotoxin; that is, they are involved in some important aspects of the diseases caused by the pathogen which produced them (Wheeler and Luke, 1963). For the purposes of this review, the term "host-specific" is reserved for those pathotoxins which exhibit effects on susceptible cultivars at concentrations at least one order of magnitude lower than the concentrations required for effects on plants designated as resistant. Host-selective toxins are defined as those which may be involved in determining pathogenesis on species or genera, but not on individual cultivars.

A report by Auriol *et al.* (1978) of a toxin from *Rhynchosporium secalis* that is host-selective cannot be accommodated readily in either of the above categories. This material is neither host-specific, (since there is no correlation with sensitivity to the preparation and varietal disease reaction) nor selective, since other genera of cereals are affected. Some resistant barley varieties appear to be affected, while some susceptible varieties are insensitive (Auriol *et al.*, 1978). Because of uncertainties as to the purity and activity of these preparations (Kono *et al.* 1981), independent confirmation of the results is essential.

II. PRODUCTION AND ASSAY

In most instances host-specific toxins have been produced in small volume (100–200 ml) still cultures. Most commonly, the medium has been Fries medium modified by inclusion of small amounts (0·1%) of yeast extract. For *A. kikuchiana* (Sugiyama *et al.*, 1965) and *A. citri* (Kohmoto *et al.*, 1979) Czapek-Dox medium has been used, but whether there were particular advantages to its use in these instances is not clear.

There has been a general belief that small-volume still culture is essential for maximum yields of host-specific toxins. However,

Sugiyama *et al.* (1965) used 200 l fermenters in order to obtain sufficient AK toxin for structural analysis. Whether actual yields per unit volume were different from those of still culture was not reported. One potential reason for the lack of success in shake or agitated cultures is the tendency for mycelium to clump, and thus perhaps provide an unfavourable interface for exchange with the medium. Van Etten and Daly (1980) recently used the technique described by Rawn and Van Etten (1978) of repeatedly blending "starter" shake cultures to train mycelium to grow homogeneously as individual small colonies. HV and T toxins were produced at concentrations equivalent to those of still cultures. The particular advantages of shake and fermenter cultures were: (a) similar amounts of toxins were produced in 5 or 6 days, compared to 3–4 weeks of still culture; and (b) more complete depletion of nutrients occurred. The latter observation may be of significance in purification of host-specific toxins. In our experience, the sometimes substantial amount of residual sugars in still cultures has been detrimental, because they have solubility properties similar to water-soluble toxic fractions.

In general, studies of production and isolation of toxins has focussed on metabolites excreted into culture fluids. In the case of T-toxins, however, it was observed in early studies that appreciable toxin was present in mycelial homogenates. Subsequent work has shown that 80–95% of highly purified T toxin arises from mycelium. Although the same toxin can be obtained from culture fluids only, structural studies of T toxin were possible only because of the large amounts extracted from mycelium. Because of the pathological implications, it is logical that culture filtrates should be considered primary sources of host-specific pathotoxins. To limit studies to extracellular production, however, may limit total yields appreciably. Further, with some organisms, secretion into the environment may be a function of cultural conditions. The presence of functional toxin thus may not be detected if only culture filtrates were examined.

Assays based on symptom production (leaf necrosis, stunting of plants, etc.) are convenient in the early phases of work on host-specific toxins; for example, in the selection of suitable cultural conditions and/or pathogenic strains. Because such assays are at best semi-quantitative, they are rarely useful in the critical stages of purification when biological activity on a unit weight basis is required. Inhibition of root growth frequently has been employed (Scheffer and Pringle, 1967) but in our experience with both T and HV toxins, this assay is not completely satisfactory even when extraordinary care (Yoder *et al.*, 1977) is taken in selecting uniform seedlings (Bhullar *et al.*, 1975). Further, determination of root inhibition requires several days. Since

Table II. Inhibition of photosynthesis in thin leaf slices of oats by crude culture filtrates from *Helminthosporium victoriae.*

Concentration ($\mu g\ ml^{-1}$)	*0*	*0·5*	*8·0*	*20*
	Park oats (Susceptible)			
Photosynthesis rate±S.D.	330±14	293±38	168±20	59±13
% Inhibition		11	49	82
	Otee Oats (Resistant)			
Photosynthesis rate	335±18	355±12	338±20	339±11
% Inhibition	—	+6[b]	+1	+1

[a] Original culture fluid was 7·2 mg ml^{-1}.
[b] Plus indicates rates above controls.

each datum point requires a minimum of 5–10 seedlings, the demands on plant materials (normally removed from commercial production because of disease susceptibility) are somewhat troublesome, especially if the seedlings must be selected for uniformity after germination. For the latter reason, we sometimes have used only 5% of the seeds initially germinated for a bioassay.

Because of these considerations, Bhullar *et al.* (1975) adapted the inhibition of dark $^{14}CO_2$ fixation by 5 mm-diameter leaf discs to use as a bioassay for T toxin. As originally described, the assay required approximately 24 h to complete. More recently, Daly and Barna (1980) used thin corn leaf slices to detect inhibition of both dark CO_2 fixation and photosynthesis in susceptible plants. Assays based on either process can be completed within 6–8 h, but photosynthesis results in much more $^{14}CO_2$ incorporation, and hence the amounts of isotope required are much less. Table II summarized some unpublished data using the thin leaf slice technique with oat leaves and HV toxin. Unlike race T toxin and corn leaves (Daly and Barna, 1980), at least 3 h or pre-incubation in HV toxin is required for significant inhibition. In the instance shown in Table II, the toxin preparation was a crude filtrate diluted approximately 10^3 to provide the indicated weights, but specificity was demonstrated clearly. When compared to oat root growth inhibition, this assay appeared to be at least one order of magnitude less sensitive. For many purposes, however, the saving in time may outweigh loss in sensitivity. Sensitivity might be increased by longer incubation times in toxin.

The use of physiological processes as assays thus has many advantages. For example, the data of Table II could be obtained (but was not) from as little as one mature oat leaf of each variety. Each replicate required only 0·5 ml of assay solution and 1 μl of toxin solution, thus conserving chemicals and space. In addition, the assays can be conducted using disposable pipettes and containers, eliminating problems of contamination (Payne *et al.*, 1980b).

The leakage of ions may serve as an assay for race T toxin (Gracen *et al.*, 1972) and HS toxin (Scheffer and Livingston, 1980). However, its use may be limited, by necessity, to relatively pure preparations essentially devoid of contaminating ions. Yoder *et al.* (1977) compared several processes in corn for their sensitivity to T toxin. The order of sensitivity was mitochondrial oxidation > dark CO_2 fixation > inhibition of root growth. The series for sensitivity may represent only differences in permeation of T toxin into tissue because the responses to concentration were not remarkably different. Certainly any or all might be useful quantitatively.

Karr *et al.* (1975) reported that T toxin could be quantitatively assayed chemically in culture fluids by the Liberman-Burchard reaction commonly employed to measure sterols. Yoder and Gracen (1977), in a careful study, showed that there was no correlation between toxin production as measured by bioassay and the development of colour by the Lieberman-Burchard reaction. The results of Karr *et al.* (1975) may have been conditioned by their belief that T toxin was a mixture of sterols and related sterol glycosides (Karr *et al.*, 1974). The results of Yoder and Gracen (1977) cast some doubt on the elaborate description of the sequence of toxin production in leaves (Karr *et al.*, 1975) which was based on determination, by the chemical method, of toxin produced in culture.

Karr *et al.* (1974) used necrosis of punctured corn leaves as an assay for T-toxin, a method similar to that used by Steiner and Strobel (1971) for HS toxin. It was stated that this was necessary because they found toxin species that inhibited root growth specificially in Tms corn but with no effect on leaves. To our knowledge there has been no confirmation of these findings. Rather, several groups have shown that assays based on the insertion of toxin into wounds on leaves are not particularly reliable or sensitive for T toxin (Bhullar *et al.*, 1975) and HS toxin (Livingston and Scheffer, 1980).

A general problem in the determination of structure of host-specific toxins is the relatively low yields that are encountered in production. This may be a particular problem with HS toxin. In the experience of one of us (J.M.D.), host-specific effects were produced by culture fluids of *H. sacchari* only at dilutions no greater than 1/1000 and most generally at dilutions of 1/5 or less. Steiner and Byther (1971) used culture fluids diluted no more than 10-fold in assaying toxin production. The problem may lie in an innately low toxicity for HS toxin or in the rate of its production.

If the rate of production is the critical factor, the difficulty apparently has been overcome with the report by Pinkerton and Strobel (1976) that serinol (2-amino-1,3-propanediol) increased production of HS toxin, even in completely attentuated cultures. It is claimed that serinol is a constituent of leaf surfaces and addition of 1 mg of leaf washings to the 10 ml of culture fluid used for each 50 ml culture resulted (Pinkerton and Strobel, 1976—Table 2) in 2·2 μmol of HS toxin per mg of dried mycelium. For the molecular weight of 236 which must be assumed from previous data (Steiner and Strobel, 1971), that amount is equivalent to 0·53 mg per mg dry weight of mycelium; i.e. approximately 33% of the total metabolic activity of these cultures was devoted to the production of an extracellular, secondary fungal product. The amount does not appear to be due to an error in calculation for that particular experiment, since the text (Page 4008) provides another value of 1·6 μmol per mg dried mycelium. Since 2·2 μmole was produced per mg dried mycelium in 5 days, and the dry weight per flask was 22 mg at 5 days, 11 mg were produced in 10 ml of culture (0·1% solution of toxin) one litre of which would contain 1·1 g of toxin. This would seem to constitute an ideal system for obtaining highly purified toxin. The yields relative to fungal growth seem improbable in view of other data on fungal toxins, but are not impossible. Because of the very rapid rate of production, successive culturing in 5–10-litre batches should supply more than sufficient toxin for the structural determinations that must be done if the proposed structure (Steiner and Strobel, 1971) is to be investigated seriously.

III. DETERMINATION OF STRUCTURE

It is axiomatic that complete understanding of host-specific toxins is dependent on knowledge of structure. Unfortunately, only three (AM, T toxins and AAl toxins) have reasonable evidence in this regard. This statement requires some explanation. Starting in the 1960s, proton nuclear magnetic resonance spectroscopy (PMR) emerged as a powerful quantitative probe for chemical structure. Subsequently, ^{13}C nuclear magnetic resonance (CMR) provided newer diagnostic aids. These developments were accompanied by vastly improved resolution of mass spectra (MS). By 1970, these tools were available for establishing the purity, as well as the composition, of natural products.

The question of purity is of particular importance in work on toxins.

The separation methods generally employed involved partitions depending on solvent solubilities and adsorption on supports and may not be highly selective, especially since biological activity, not chemical functions, are monitored. The very high biological activities of host-specific toxins often may be misleading in the estimation of purity. Obtaining a preparation with biological activity of 10 ng ml^{-1} in a root assay, for example, can be exhilarating but should be tempered by the realization that, if the toxic entity itself has an intrinsic activity of only 1 ng ml^{-1}, even such an active preparation contains 90% of impurities.

PRM, CMR and MS data can resolve some of these problems. In combination, these techniques should permit unambiguous structural assignments if a compound is pure. High resolution MS, in particular, provides exact elemental compositions and molecular weights. PMR and CMR should provide spectra in which all absorptions can be accounted for in terms of a pure compound. Absorptions which cannot reasonably be assigned raise questions about purity.

For these reasons, the discussion on structure is developed chronologically. The techniques readily available before 1970 permitted certain types of proofs which were acceptable at that time, but which today are not considered adequate. Of course, the ultimate proof of structure is laboratory synthesis of biologically active material that reproduces the specificity and toxicity of the natural compound.

A. HV Toxin

The first attempts at determination of structure were those of Pringle and Braun (1957) who, by chromatographic procedures, prepared a toxic ninhydrin-negative fraction with a specific activity of 10 ng/ml in the root growth inhibition assay. Pringle and Braun (1955) noted that treatment with $NaHCO_3$ resulted in loss of activity concomitant with the appearance of two ninhydrin-positive components upon paper chromatography. One component was subsequently identified as a nitrogenous base (victoxinine, $C_{17}H_{29}NO$). It was assumed that HV toxin was a cyclic polypeptide covalently linked to victoxinine.

Unfortunately, there has been no further substantial work on HV-toxin since that time, and the amount of useful physical and chemical data obtained on that preparation was limited. It was reported that a major problem to progress was instability when toxin was purified (Pringle and Braun, 1957), but the conditions for, and the nature of, the instability were not detailed.

A later study by Pringle and Scheffer (1963) reported additional chromatographic procedures which resulted in a specific activity of 0·2 ng/ml in the same bioassay used by Pringle and Braun (1957). No analytical results were reported on the more active preparation of Pringle and Scheffer (1963). As suggested by Pringle and Scheffer (1963), the 50-fold increase in activity may be the result of removing inactive (denatured?) toxin of the same chemical species, but there are problems in this reasoning (Kono *et al.*, 1981). It remains possible that the toxin preparation of Pringle and Braun (1957) upon which analysis was performed was contaminated by species of peptides.

B. PC Toxin

Pringle and Scheffer (1963–1966) first reported a crystalline product for PC-toxin which affected susceptible sorghum cultivars at 0·1–1·0 μg ml^{-1} in root assays. Later they reported the existence of at least two toxin entities which apparently co-crystallize (Pringle and Scheffer, 1967). As with HV-toxin, evidence for a peptide structure was obtained by hydrolysis-yielding alanine, aspartic acid, glutamic acid and serine in the ratio of 3:3:7:1. The amino acids accounted for approximately 75% of the weight of the crystals. Molecular weight was estimated at 2000 daltons by molecular sizing.

More recently, Wolper and Dunkle (1980) also have reported two toxin moieties for PC-toxin. They report activity at 1 ng ml^{-1}, although the assay endpoint (50% inhibition of root growth) is different from that used by Pringle and Scheffer (1963) (100% inhibition of root growth). Unlike Pringle and Scheffer (1967), their preparation is stable and, surprisingly, yields only one amino acid, aspartate. Although further work is required, the instability of earlier preparations and the presence of four amino acids might be due to the presence of contaminants, perhaps a crystalline material with adhering toxin.

C. HC Toxin

A crystalline preparation also was reported for HC-toxin (Pringle and Scheffer, 1967) with activity at 0·5 μg ml^{-1}. Again the evidence suggested a cyclic peptide of at least 5 amino acids, one of which was an α, β unsaturated derivative of isoleucine (Pringle, 1971). A puzzling aspect is a reported molecular weight of less than 700 daltons by molecular sizing. The molar ratios obtained for the component amino acids suggests a minimum molecular weight of 2400 (Kono *et al.*, 1981). Further, a non-crystalline preparation obtained by Yoder and Scheffer (1973) showed somewhat higher toxicity (0·2 μg ml^{-1} for 50% inhibition of root growth), but this may be a result of variation in sensitivity of seed lots.

D. HS Toxin

In 1971, Steiner and Strobel reported characterization of HS toxin as a cyclopropanol galactoside with a molecular weight of 236. The evidence for this structure has been discussed in detail by Kono *et al.* (1981) and questions about its validity will be only briefly summarized here. The sensitivity and reliability of the bioassay has been questioned previously (Bhullar *et al.*, 1975; Wheeler, 1976; Daly, 1976). Although the concentration curves for biological activity are expressed in terms of 10^{-8}, 10^{-9} etc. mol and thus imply a very active molecule, these terms are misleading. Such amounts were routinely applied to leaves in 1 μl (10^{-6}) of solvent. Thus, active solutions were at concentrations of 10^{-2} to 10^{-4} M (2·36 mg ml^{-1} to 236 μg ml^{-1}), unremarkable concentrations for a presumably biologically potent compound. Since this preparation appeared to contain galactose, there is the complication that galactose is an unusual carbohydrate in its toxicity to higher plants, including sugar cane cell cultures at 5×10^{-3} M (Maretzji and Thom, 1978). There are a number of observations on galactose toxicity (some of these are cited by Colcasure and Yapp, 1976) extending back to Knudson (1915). This literature has been overlooked, but it has a bearing on toxic activity of culture filtrates.

The evidence for purity of HS toxin preparations by Strobel's laboratory is not overwhelming, being largely dependent on the fact that a single spot, detected by chemical staining, was obtained with paper and thin layer chromatography (TLC). The NMR spectra contained at least one peak that could not be assigned (Steiner and Strobel, 1971). Wheeler (1976) also has pointed out, from existing biological evidence, the possibility that the toxin entity may constitute as little as 5% of the weight of these preparations.

Even if purity is assumed, the evidence for the proposed structure is far from satisfactory. Cyclopropane bonds generally are unstable in alkali or acid, and it is difficult to understand how the proposed structure could persist intact through the final two steps of purification in solvents containing 30% NH_4OH. The largest mass in MS is 236, which is in agreement with the proposed structure (Steiner and Strobel, 1971), but the spectrum is of low resolution. Thus, there is no evidence that the largest peak is in fact the molecular ion, because its molecular formula was not determined. The same considerations apply to the identification of the mass 57 and 73 peaks as the cyclopropanol aglycone with and without glycosidic oxygen. There are a large number of compounds of C, H and O which these three masses might represent, including expected fragments from galactose. A major problem in mass spectroscopy is lack of volatility of many compounds without the use of probe temperatures which lead to excessive thermal degradation. Our laboratory encountered this problem in work on T toxin (Kono and Daly, 1979; Kono *et al.*, 1980). It is not improbable that the m/e fragment of 236 represents only a part of a larger molecule. Although examination of the isotopic masses sometimes can resolve these difficulties (Silverstein *et al.*, 1974), it was not, or could not be, attempted in this instance.

The PMR spectra which should provide substantial evidence for correctness of the proposed structure is poorly resolved, but perhaps because of limiting amounts of material. As pointed out by Kono *et al.* (1981a), the infrared spectra indicate large amounts of water in the preparations, some of which may be organically bound. In any event, the necessary quantitative examination of protons associated with specific functional groups is difficult to undertake from the published spectra, although values are presented in the text.

Some of the PMR assignments are quite dubious. For example, assignments for the protons at the C_5 and C_6 positions of the sugar moiety were made, but to our knowledge similar assignments have not been made for the numerous published spectra of mono- or disaccharides, nor for glycosides such as that proposed by Steiner and Strobel (1971). The precedent for these assignments was cited as a paper of Pasika and Cragg (1963). Pasika and Cragg made these assignments for isomaltotriose, a trisaccharide with two C_1-O-C_6 linkages. The deshielding effect (Silverstein *et al.*, 1974) of this linkage results in a spectral shift which can be identified. However, since there is no such linkage in the proposed structure of HS toxin, it is difficult to understand how the assignment can be made for it, but not other glycosides, unless additional techniques are used. A major problem with the published HS toxin spectra is the use of HDO as a solvent, since the HDO peak occurs in a region of the spectrum which obliterates some diagnostic carbohydrate absorptions. Solvents such as pyridine or demethylsulphoxide would be more appropriate (Kono *et al.*, 1981).

Similar uncertainties prevail for the PMR assignments for the cyclopropane protons. Spectra of cyclopropane compounds are limited in number, and we could find none with the dioxo functions proposed. In general, the assignments made by Steiner and Strobel (1971) are not close to possible reference points. For example, the assignment of methylene protons to absorption at $\delta 1{\cdot}35$–$1{\cdot}60$ and $\delta 1{\cdot}2$–$1{\cdot}35$ is quite far from the absorptions at $\delta 0{\cdot}68$–$0{\cdot}74$ and $\delta 0{\cdot}65$–$0{\cdot}81$ for phenylcyclopropane and 1-methyl-1-phenylcyclopropane in references which they cite as support for the structure.

Although it seems likely that galactose was in these preparations, in view of the above uncertainties it seems unwise to apply the trivial name Helminthosporoside to them. The abbreviation HS toxin is more appropriate, and for several reasons. If the organism eventually is placed in the genus *Bipolaris*, rather than *Helminthosporium*, the abbreviation BS toxin preparation remains suitable.

Recently, R. P. Scheffer made available to us a preprint by Livingstone and Scheffer (1981). Because they found the leaf puncture bioassay (Steiner and Strobel, 1971) unreliable, they monitored ion leakage from roots of susceptible plants during purification. The final preparation had activity at 1–10 ng ml^{-1}, and thus appears to be much more active than the HS toxin preparation of Steiner and Strobel. Whether it is completely pure remains to be determined. The available MS and PMR spectra are not definitive on this point. Like Steiner and Strobel, they present evidence for the presence of galactose, but as a tetra- or penta-saccharide with an unusual 1,5 glycosidic bond. More important, there is evidence for a C_{15} aglycone moiety which may be a sesquiterpene derivative.

E. AK Toxin

The presence of phytotoxic compounds in cultures of *A. kikuchiana* has been recognized for some years. Sugiyama *et al.* (1965, 1966), isolated and identified a toxic compound and gave it the trivial name alternin. The structure has been confirmed by synthesis (Sugiyama, 1967). As far as we are aware, host-specificity for the purified natural product or the synthetic compound has not been demonstrated.

Two separate groups have isolated host-specific toxins. Ohkawa and Tanekata (1967a, 1967b) found a component (AK-III?) which did not contain nitrogen, halogen or sulphur. Otani *et al.* (1972, 1973) apparently isolated a single host-specific toxin (AK toxin), but Nishimura *et al.* (1979) indicate there are at least three host-specific toxins in *A. kikuchiana* cultures. The picture is further confused by the presence of known non-specific toxins (Nishimura *et al.*, 1979). In any event, details concerning the structure of the host-specific components have not been obtained.

F. AM Toxin(s)

The work on structure of toxin from *Alternaria mali* is most satisfying because two groups independently arrived at identical structures for the major host-specific toxin. At nearly the same time, Okuno *et al.* (1974a,

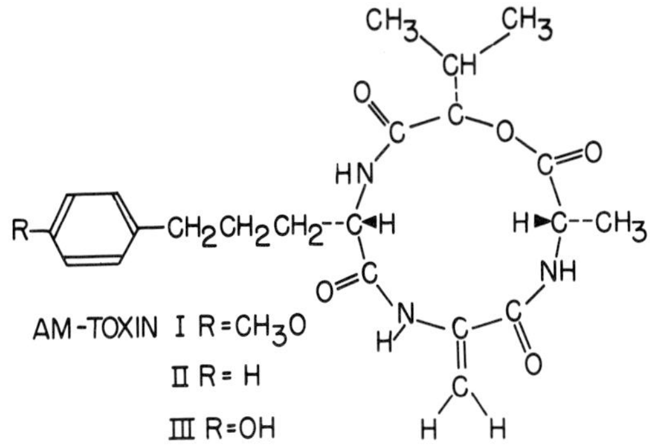

Fig. 1. Structure of host-specific toxin(s) from *Alternaria mali* independently obtained by Ueno *et al.* (1975) and Okuno *et al.* (1975).

1974b and 1975) and Ueno *et al.* (1975a, 1975b, 1976) determined the structure of AM I-toxin shown in Fig. 1. The preparation of Ueno *et al.* (1975a) showed a specified activity at 0·2 ng ml^{-1}. Ueno *et al.* (1976) also provided evidence for the structures of AM II- and AM III-toxins (Fig. 1). The similarity to the non-host specific toxin tentoxin (Myer *et al.*, 1971) should be noted. Tentoxin also is found in cultures of *A. mali.*

It seems unnecessary to detail the analytical proofs (See Kono *et al.*, 1981) because laboratory synthesis has been accomplished (Lee *et al.*, 1976). The achievements of these groups of research workers represent a landmark in the field of plant pathology.

G. T Toxin

Karr *et al.* (1974) reported that T toxin consisted of four components which were suggested to be steroid-like terpenoids and their glycosides. The evidence appears to have been only the fact that ^{14}C-mevalonate apparently was incorporated into their preparations. The NMR spectrum does not show enough detail to confirm the suggestion. Two of the toxin fractions gave identical molecular ions at m/e 388·2625. Aranda *et al.* (1978) proposed that T toxin consists of D-mannitol esterified by acetic acid and by two-N-formyl-L-valine moieties. The analytical data in this paper are minimal, and the structure is at variance with the lack of nitrogen in the preparation of Karr *et al.* (1974).

Kono and Daly (1979) also reported a toxin preparation devoid of N (C:63·84; H:9·18; Q:26·68) with biological activity in the ng ml^{-1} range (10^{-8} to 10^{-9} M) (Payne *et al.*, 1980a). In contrast, Berville (1978)

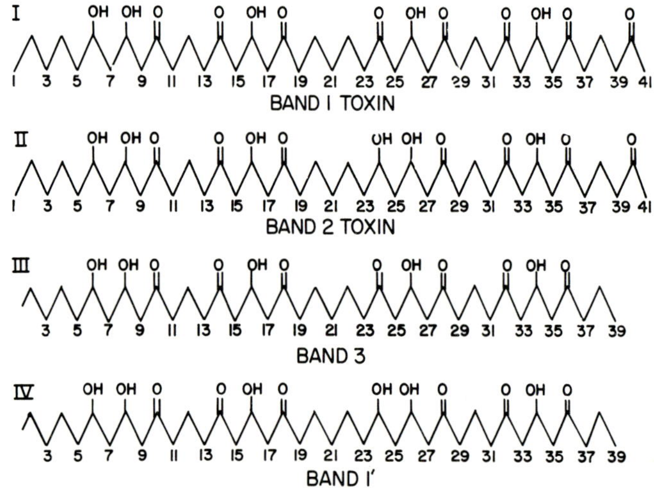

Fig. 2. Structure of four of the major components from the T toxin complex.

found that the preparation of Aranda *et al.* (1978) was active only in the μg ml^{-1} range. First isolated in 1976 in our laboratory, we believed for two years that the toxin was a single compound of empirical formula $C_{41}H_{68}O_{13}$ and with the general structure shown as band 1 toxin in Fig. 3. However, a final conclusion could not be drawn because of inability to obtain a m/e of 768 in mass spectra, as the empirical formula demanded. The highest mass observed in high resulation MS had an empirical formula of $C_{33}H_{44}O_5$, indicating considerable pyrolysis and degradation during MS anlaysis.

It perhaps was fortunate that these problems arose because two subsequent lines of evidence suggested that the preparation was actually a mixture of very closely related compounds. In an attempt to verify that the molecule was linear the toxin preparation was reduced completely to hydrocarbons and analysed by mass spectroscopy. Instead of a single hydrocarbon ($C_{41}H_{84}$), a series ranging from C_{35} to C_{45} were obtained (Kono and Daly, 1980). However, the C_{41} hydrocarbon predominated, and it was not possible to decide if the unexpected compounds were natural products or compounds arising by rearrangement during reduction.

Independently, Ronald Betts and Carl Tipton (personal communication) had obtained by different methods a T toxin preparation that appeared, chromatographically and chemically, to be identical to ours. However, in a systematic examination of a number of TLC supports and solvents, they found one (and only one) commercial plate (Merk EM 60 Silica gel), which provided 7–9 distinct spots. The possibility of artifacts

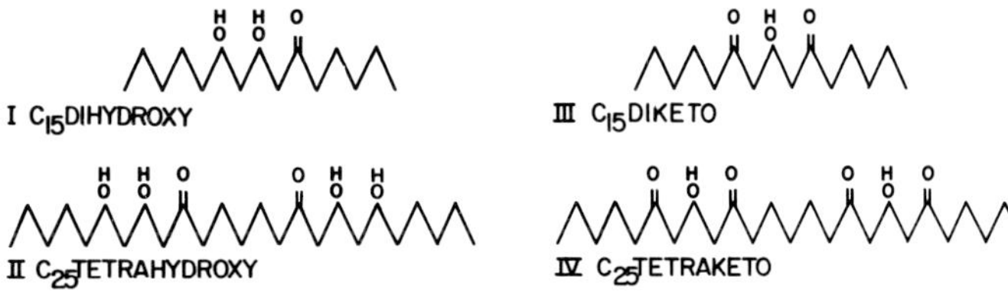

Fig. 3. Some compounds synthesized by Y. Suzuki in our laboratory which show host-specific effects on Texas male sterile cytoplasm corn tissue.

due to chromatography could not be eliminated, however, and it was necessary to isolate individual components even though most were in low yield.

Each component was host-specific and toxic to the same extent within error (Payne *et al.*, 1980a). More important, the specific biologic activity of the original toxin preparation was equivalent to each of three individual components isolated in amounts sufficient to test in detail, thus indicating that all components of the mixture were equally active. By a slight modification of sample introduction into the mass spectrometer, the trimethyl silyl esters of the four major components were shown to have empirical formulae of $C_{41}H_{68}O_{13}$, $C_{41}H_{70}O_{13}$, $C_{39}H_{66}O_{12}$ and $C_{39}H_{68}O_{12}$. More recently, the new technique of field desorption mass spectrometry has confirmed the presence of these four compounds as well as additional minor C_{37}, C_{43}, C_{45} components in the original, unchromatographed, underiviatized toxin Kono *et al.*, 1981a, b). The elemental compositions of the minor toxins are quite similar to the major components (Kono *et al.*, 1981b).

In the first paper of this series (Kono and Daly, 1979), it was postulated from chemical reactions, CMR, PMR (with decoupling) and MS that the two major components (both C_{41}) had the structures indicated as Bands 1 and 2 toxins (Fig. 2). However, it was not possible to determine the exact positions of the oxygen groups. Kono *et al.* (1980) have confirmed the positions shown in Fig. 3 in an unambiguous analysis of mono-oxygen derivatives and from the details of the CMR spectra of the individual underiviatized toxins species.

Kono *et al.* (1980a, b) show that the C_{39} components are counterparts of the C^{41} moieties, differing by the absence of a terminal methyl ketone group. This finding has bearing on biosynthetic pathways as well as on the mode of action of T toxin. Since band 1 and band 3 toxins (Fig. 2) have equal toxicity (Payne *et al.*, 1980a), the terminal methyl ketone function (Fig. 3) appears to be of no consequence in either toxicity or host-specificity.

The above discussion has not focussed directly on the experimental details of the proof of structure for several reasons. More complete descriptions are given by Kono *et al.* (1981a). It should be noted that: (a) IR, CMR, PRM and MS spectra show no absorptions or peaks that cannot be accounted for by the proposed structures; (b) in the case of PMR spectra, quantitation of protons is in good agreement with expected values; (c) all assignments for CMR and PMR have strong precedents in existing literature; and (d) the formation of derivatives such as methyl esters, acetates or trimethylsilyl ethers yield the expected analytical spectra.

Despite this evidence for purity, it can be argued that these compounds are not host-specific toxins, but show toxicity only because a very active toxin is carried along as an impurity. This argument is considerably weakened with the synthesis by Dr Y. Suzuki in our laboratory of the structures shown in Fig. 3. Although the chain lengths are not as long as native toxin, each produces inhibition of photosynthesis and dark CO_2 fixation in leaf slices of Texas male sterile corn, but not in corn with normal cytoplasm even at very high (40–60 $\mu g\ ml^{-1}$) concentrations. Preliminary studies indicate that the order of toxicity appears to be: $I < III < II < IV \cong$ native toxin. Compound IV is only slightly less active, if at all, than native toxin isolated from *H. maydis*, race T. If our preparations from the fungus contain an undetected toxin, it also has been produced unknowingly by laboratory procedures designed to synthesize model compounds. The fact that synthetic compounds have the biological activities of the natural toxin supports the analytical work on which the structures shown in Fig. 2 depend.

H. AAl Toxin

Gilchrist and Grogan (1976) described host-specific necrosis of Early Pak tomato caused by culture filtrates of an isolate of *A. alternata* f. sp *lycopersici* virulent on this variety. The culture filtrates subsequently yielded a ninhydrin-positive fraction which could be resolved by gel-permeation chromatography, isoelectric focussing, and TLC into two apparently homogenous toxin components, T_A and T_B (Bottini *et al.*, 1981). In preliminary work, however, NMR spectra indicated that each component consisted of two molecular species. This problem just recently has been resolved by an elegant example of NMR analysis.

Hydrolysis of component T_A of AAl toxin yielded the salt of 1,2,3-propanetricarboxylic (=tricarballylic) acid and a ninhydrin-positive fraction whose molecular formula was found to be $C_{19}H_{41}NO_5$ by high resolution MS of its methylated and deuteriomethylated

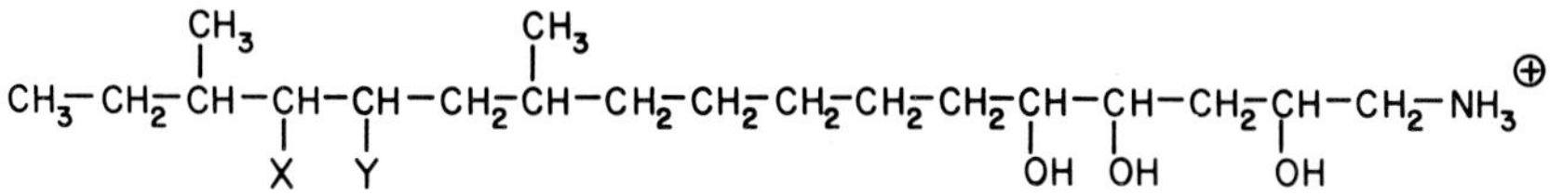

1·H^+ = **1a**: X = Y = OH

2a: X = OH, Y = $^{\ominus}O_2C{-}CH_2{-}CH(CO_2^{\ominus}){-}CH_2{-}CO_2$

2b: X = $^{\ominus}O_2C{-}CH_2{-}CH(CO_2^{\ominus}){-}CH_2{-}CO_2$, Y = OH

Fig. 4. Structural assignments for two host-specific components of fraction T_A obtained from *A. alternata* f. sp. *lycopersici*. Figure provided through the courtesy of A. Bottini and D. G. Gilchrist. Two host-specific components of fraction T_B appear similar except for lack of hydroxyl on C_5.

derivatives (Bottini and Gilchrist, 1981). MS fragmentation also revealed the presence of 5 hydroxyls and a primary amine group. PMR and CMR spectra led eventually to the functional group assignments shown in Fig. 4, compound 1a (1-amino- 11,15 dimethylheptadeca -2,4,5,13,14-pentol), although it was not at first possible (Bottini and Gilchrist, 1981) to distinguish between 11,15- or 7,15-dimethyl isomers.

The presence of 5 hydroxyls in this unusual aminopentol (Fig. 4) posed considerable difficulty in establishing the site or sites of esterification of tricarballylic acid. This was accomplished by analysis of the CMR spectra of the mixture of species comprising fraction T_A (Bottini *et al.*, 1981). In addition to absorptions corresponding to carbonyl groups, there were twelve which were assigned to a major component, twelve corresponding to a minor component, and seven which were nearly identical to seven signals present in the free aminopentol. The latter were assumed to be common in both species. Four of these bands were identified as bands assigned to, and calculated for, C_1, C_2. C_4 and C_5 of the free aminopentol (Fig. 3); therefore esterification must occur at C_{13} and/or C_{14}. Further, the other three common ^{13}C signals were sufficiently close to those calculated for the methylene carbons at C_3, C_6 and C_7 so as to lead to the choice of the 11,15-dimethyl isomer (Bottini *et al.*, 1981), rather than the 7,15 dimethyl isomer.

Calculation of the ^{13}C chemical shifts expected for acylation of either C_{13} or C_{14} and comparison with observed chemicals shifts led to the conclusion that the major component is esterified with tricarballylic acid at the C_{13} position while the minor component is esterified at C_{14}. In addition, a terminal carboxyl of the acid is involved, not the carboxyl at C_2 (Bottini *et al.*, 1981). The two components of T_A generally are found in a 4:3 ratio. Research in progress suggests that the second toxic fraction isolated by TLC (T_B) also consists of two components differing by lack of the hydroxyl at C_5 and in the stereochemistry of groups from C_{11} to C_{15} (Bottini *et al.*, 1981). Bioassay (Gilchrist and Bottini, unpublished) so far has failed to distinguish differences in innate toxicity between T_A or T_B or the two esters of T_A. Bioassay also shows that the free aminopentol and the tricarballylic acid are very much less active than fraction T_A (Bottini and Gilchrist, 1981) and, presumably, each of the four individual toxin components.

This outstanding structural analysis augurs well for future research by the group at the University of California, Davis, for several reasons. As discussed later, AAl toxin shows significant differences in its biological effects from most other host-specific toxins. With knowledge of the structure, and the presence of the single locus controlling host toxin sensitivity, this system well may provide the first clear chemical explanation of specificity in host-parasite interactions.

I. Other Toxins

The amount of analytical work published for the other toxins given in Table I is not yet sufficient to justify extensive comment at this time. The T-like toxin of *Phyllosticta maydis*, discovered independently by Comstock *et al.* (1973) and Yoder (1973), is of some theoretical importance because it is the only known case for fungi in different genera producing toxin of the same host specificity. In our experience, and that of others (Yoder, personal communication), yields from *P. maydis* are much lower than from *H. maydis* race T.

It is claimed that *Rhynchosporium secalis* produces a "host selective" toxin (Auriol *et al.*, 1978). The claim is not listed in Table I because its biological basis is not clear. There is no correlation of toxin activity with susceptibility to *R. secalis*. Some varieties resistant to the pathogen were sensitive to the toxic preparation while some susceptible varieties were insensitive in the host species barley and rye. There were similar findings even for the non-host species wheat and oats (Auriol *et al.*, 1978). It is difficult to ascertain the concentrations required. The bioassay section of material and methods indicates tests were run at 25 mg ml^{-1} while later in the text it is indicated as 25 μg ml^{-1} (Auriol *et al.*, 1978).

Even if claims for host selectivity are shown to be valid, problems exist in the acceptance of the proposed structure for this material as a cellobioside of 1,2 propanediol. One difficulty lies in the data given for the MS of the aglycone. They do not correspond to data published in the EPA-NIH sourcebook of mass spectra. The second difficulty arises in the quantitative aspects of the published NMR. Although most of the assignments are not unreasonable, they yield fewer numbers of protons than claimed (Auriol *et al.*, 1978) or expected from theory. For example, the protons for the pyranose rings in cellobioside at $\delta 3{\cdot}40$ to $\delta 4{\cdot}08$ ppm should integrate for 15; 14 are claimed (Auriol *et al.*, 1978). We obtained only 10 from the published spectra (see Kono *et al.*, 1981 for details). Similarly, there should be two anomeric protons in this compound; our analysis yielded slightly less than one. The latter result would be expected for a non-glycosidic, free carbohydrate. There are strong similarities in the published spectra and those known for methyl pentoses, such as rhamnose and fucose, which frequently are found as fungal wall components. Although it might be argued that the published spectrum is not typical for the preparation, only spectra that prove the essential points should be presented.

Very recently, Pio Beltran *et al.* (1980) have revised the earlier conclusion and now suggest that the material is a mixture of higher oligosaccharides of 1,2 propanediol. No physical evidence was presented, except that the R_f values of the material isolated from cultures are different from synthetic (2 – 0) α-1,2-propanediol cellobioside. A toxic radioactive trisaccharide was claimed to have been synthesized from (1 – 0) α-1,3-propandiol glucoside and ^{14}C-glucose through the reverse action of β-glucosidase. There are several aspects of the experiment which are inconsistent with the expected results (see Kono *et al.*, 1981). The major discrepancy lies in the fact that the described protocol should yield a product with a specific radioactivity of $1{\cdot}4 \times 10^6$ dpm mg^{-1}. The reported value was 3×10^4 dpm mg^{-1}. There is no ready explanation for a radiochemical yield of approximately 2% of the theoretical yield. If, in fact, the material is a trisaccharide glycoside, it is difficult to reconcile the published NMR spectrum (Auriol *et al.*, 1980) even with the stated proton integration. There should be 21 protons at $\delta 3{\cdot}4$ to $\delta 4{\cdot}08$ PPM in the pyranose rings of a tricelliobioside, not 14 (Auriol *et al.*, 1980).

The group at Montana has also reported two toxins of *A. solani* which act only synergistically to produce host-specific activity against potato (Mattern *et al.*, 1978). It was suggested that both compounds were lipid-like because they apparently could be detected by phosphoric acid and anisaldehyde sprays, but neither is doagnostic for lipds. Recently, however, it has been reported that the procedures for isolation could not be repeated (Shephard *et al.*, 1980).

J. Summary

The early work on host-specific toxins was inconclusive, but pointed to peptides as the active agents in three instances. It is a general principle that peptide structure is responsible for specificity in basic biochemical processes. Thus, the initial results seemed reasonable, but, so far have not been conclusively demonstrated. The only proven case of peptide linkages is for AM toxin, while the toxin from *H. maydis* race T is not a peptide. Further, toxins from the closely related *H. sacchari* appeared to be devoid of nitrogen (Steiner and Strobel, 1971; Livingston and Scheffer, 1980). One of the AK toxins also appears to be devoid of nitrogen and partially purified toxins from *A. citri* did not react with ninhydrin (Kohomoto *et al.*, 1980). Nitrogen determinations apparently were not made. In a paper to be published, Bottini *et al.* (1981) did not find any amino acids in the host-specific toxin of *A. alternata* f.sp. *lycopersici* affecting tomato.

For those toxins which apparently have no peptide bonds, the biochemical basis for their exceptional specificity for certain crop varieties may be difficult to identify.

IV. MODES OF ACTION

A. Some Perspectives

The initial point of research departure for host-specific toxins is the reproduction by fungal products of some characteristic visual symptom of the disease. At the point when molecular modes of action are contemplated, visual symptoms are of little value. Unlike more biochemically specialized animal species, plants respond to toxins in a limited number of ways. Visual symptoms of necrosis, chlorosis, growth distortions and wilt can all result from damage to quite diverse cellular biochemical processes.

1. Primary and Secondary Effects

The above comments also apply to relatively general physiological or biochemical symptoms such as changes in ion balance or respiration. The first effect has been widely used and, as discussed later, has tended to focus attention on plasma membranes as a site of action for host-specific toxins. It is obvious, however, that leakage of ions may be a reflection only of damage to a cytoplasmic function. A particular problem with host-specific toxins is that several apparently fundamental processes, such as ion leakage or increases in respiration, occur much more quickly

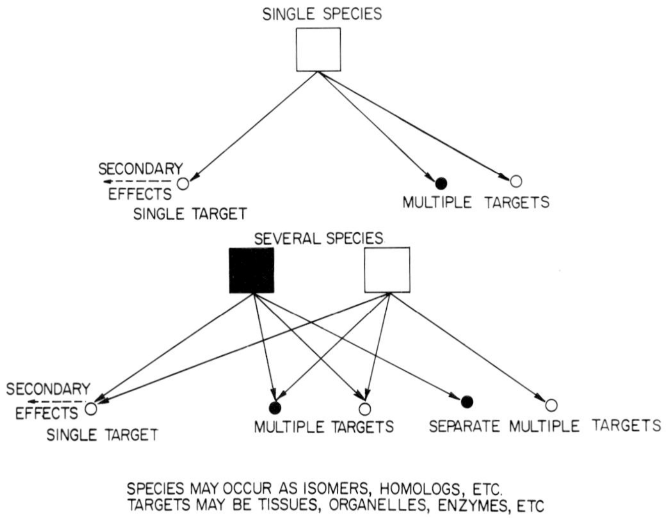

Fig. 5. Theoretical considerations which arise from the demonstration of multiple host-specific toxins.

than do other symptoms, sometimes within minutes (Wheeler and Black, 1963; Payne *et al.*, 1980a). Because of the marked specificity of this class of phytotoxins, it is logical to assume that a single target site has been affected, but with important secondary effects. Less often considered, but a possibility that must be acknowledged, is that there may be separate multiple targets (Fig. 5, upper section).

With the discoveries (see Section III) that several chemical species with the same specific toxicity often are produced, a third possibility exists. Each individual toxin moiety may have distinct and separate cellular targets (Fig. 5, lower section). Because of the similarities among the individual AM (Fig. 1) and T toxin components (Figs 1 and 2) this would be surprising, but not impossible to consider until further data on the relationship between structure and activity are obtained.

To sort primary from secondary effects, two experimental considerations generally predominate. It is logical to assume that either concentration or time dependency differences should distinguish them. Early work led Black and Wheeler (1966), Wheeler and Hanchey (1966), Pringle and Scheffer (1964) and Samaddar and Scheffer (1968) to conclude that the rise in respiration found in disease and also caused by host-specific toxins was secondary to damage to the plasmalemma because ion leakage occurred most rapidly.

Neither separation in time or differences in response caused by concentrations difference is foolproof, however. At least two elements are involved: the normal flux of metabolites that are being affected and the sensitivity of the instrumentation used to measure them. If ion flux normally is near zero, for example, a response to toxin may be much easier to detect (Payne *et al.*, 1980a) than an effect on respiration. Respiration normally involves substantial fluxes of gases and a change in the large normal flux may be difficult to detect at early time periods or at low concentrations of toxicant. Further, not all gas exchange is due to mitochondrial oxidation. Some arises from processes (CO_2 evolution during synthesis of pentoses) that may not be affected by toxin.

2. The Use of Physical Techniques

In view of the limited information that current approaches have provided with respect to molecular modes of action, it seems appropriate to comment on the future use of isolated organelles in the search for target sites for host-specific toxins. Although much remains to be done, techniques for isolation of plasma and vacuolar membranes are improving rapidly. The elegant work of Steele *et al.* (1976, 1978) has clearly shown that coupling factor 1 (CF1) in chloroplasts is the site for the action of tentoxin. This work was dependent on prior knowledge of its structure by Meyers *et al.* (1971), the early work by Templeton (1972) and associates that suggested light-dependent reactions were responsible for symptom development, and the demonstrations by Arntzen (1972) and Bennet (1976), that photophosphorylation was affected. Finally, the precise determination of binding constants by Steele *et al.* (1976) required tentoxin of high specific radioactivity, possible only after Rich and Mathiapuraman (1974) provided laboratory routes for synthesis of tentoxin.

The determination of target sites by studies of binding kinetics for phytotoxins thus has the minimum requirements of pure toxin, and labelled in such a way as to provide a sensitive assay. It is most convenient to use radioactivity as a marker but, in view of the small amounts of material required for toxicity, it must be of very high specific activity. Steele *et al.* (1976), for example, used tentoxin with $1{\cdot}2 \times 10^5$ dpm per μg in establishing a binding affinity of 10^8 M for CF1 particles and for chloroplasts.

At present, the unknown purity of most host-specific toxins precludes application of these techniques even when radioactive label can be incorporated through biosynthesis. The report of Strobel (1973) on binding of HS toxin to protein suffers in this respect (see Section III, D) as does work of Ireland and Strobel (1978) on binding of T toxin to cell homogenate.

Even if the purity of toxins in the work of Strobel (1973) and of Ireland and Strobel (1977) is unquestioned, there remains the substantial point as to whether the specific radioactivity employed in both studies would be high enough to detect binding to the naturally occurring target. Neither study made correction for the non-specific binding which occurs in studies with hormones (Hertel *et al.*, 1972) or with Fusicoccin (Dohrmann *et al.* 1977).

In the case of T toxin (Ireland and Strobel, 1977), an affinity constant of 10^{-4} M is claimed, yet other studies show (Payne *et al.* 1980a, 1980b) that purified toxin is active at 10 ng ml^{-1} or less. This would constitute a concentration of 10^{-8} M for molecular weight of 768 daltons (Kono and Daly, 1980; Kono *et al.*, 1981a, 1981b), some 10 000-fold less than the affinity constant value reported by Ireland and Strobel.

Recently, Livington and Sheffer (1981) reported the same concentration ranges (ng ml^{-1}) for effective concentrations of HS toxin against sugar cane. Although the molecular weight is not known for certainty, if the molecular weight postulated by Steiner and Strobel (1971) is used this would be equal to approximately 10^{-8} M, much smaller than the binding constant claimed by Strobel (1973).

3. *The Use of Non-host Species*

It is understandable that research on the action of host-specific toxins should focus chiefly on the responses of susceptible tissues. The conceptual yields from this approach have been small, and it may be pertinent that knowledge of the molecular action of Fusicoccin (FC) has been developed on non-host species, not on peach or almond tissue in which wilting symptoms first called attention to the existence of toxin.

There are several reasons why non-host species should be used more extensively. The effects of host-specific toxins are rapidly lethal for susceptible tissue. Consequently, the metabolism observed even in short time periods may represent only non-specific death throes from which the crucial, initial and specific events may be sorted only with considerable difficulty. In contrast, the detailed understanding of Fusicoccin was possible because FC caused growth stimulation in non-host species. In turn, these events could be related to a vast body of literature on mechanisms of hormonal action (Marre, 1979). Further, the effects could be probed experimentally in a much longer time frame than wilting phenomena.

Fusicoccin may not be an isolated example. Helminthosporal (De Mayo *et al.*, 1961) and Helminthosporol (Hashimoto *et al.*, 1967) differ only in the reduction of a carbonyl group to alcohol, yet the one causes growth effects, not necrosis, in non-host species while the other has not been reported to do so. Cronatine also causes necrosis in host tissues, but induces cellular proliferation in non-host potato tubers (Sakai *et al.*, 1979). In such instances, perhaps, similar primary targets may be affected but the secondary consequences are different. As a speculative example, it is believed that the primary action of FC is to promote H^+ extrusion by ATPase activation in some species (Marre, 1979). In turn, proton extrusion causes "acid" promotion of growth but cytoplasmic pH balance is maintained by internal processes. In other tissue pH compensation may not occur and necrosis will follow, rather than growth.

A more subtle approach for host-specific toxins may be increased emphasis on the effects caused in resistant host tissue. In our opinion, the examination of resistant host tissue has been too cursory. The impression received is that resistant tissue is examined only to prove the specificity of the effects on susceptible tissues. If effects are seen on resistant tissue at high concentration, there is a tendency to assume that the results are due to impurities. The first consequence of these research approaches is to ignore potentially important leads for mode of action. The second consequence is that host-specific effects are considered to be none (resistance) or all (susceptible) effects. This leads directly to the tenet that specificity is due to the absence or presence of a specific site or process in the host.

It has been generally assumed, for example, that only mitochondria from Tms corn are affected by T toxin. The available evidence led Watrud *et al.* (1975) to postulate that Tms corn mitochondria differed from N corn mitochondria in a property of the outer membrane which permitted access of T toxin to sensitive sites. Removal of outer membranes apparently caused normally resistant mitochondria to respond to toxin (Watrud *et al.*, 1975). Payne *et al.* (1980b) found inhibition of malate oxidation of resistant mitochondria at high (μg ml^{-1}) concentrations of toxin and also with an acetate derivative. Interestingly, mitochondria from non-host soybean plants responded qualitatively and quantitatively as mitochondria from resistant corn. Since malate oxidation occurs in the inner membrane, toxin must be penetrating even in resistant mitochondria, but still T toxin showed selective or specific host effects on NADH and succinate metabolism. The experiments of Payne *et al.* (1980b) on resistant mitochondria would not have been performed except for the assurance of pure toxin.

B. Postulated Mechanisms

It is not possible to discuss molecular mechanisms of action with a completely satisfactory organization. The following is based on presumed cellular sites of action, although few specific mechanisms (i.e. enzyme inhibition) have been identified.

1. Plasmalemma

The extensive literature of ion leakage caused by host-specific toxins has directed considerable attention to this organelle, especially in the case of victorin. Hanchey and Wheeler (1979) have provided a comprehensive review of evidence on the role of membranes in plant disease. The early evidence was indirect (Wheeler and Black, 1962) and consisted of the demonstration that ion leakage developed much earlier than did an increase in respiration. The lack of a direct effect of HV toxin on isolated mitochondria (Black and Wheeler, 1966; Wheeler and Hanchey, 1966) also tended to minimize the role of oxidative metabolism in the disturbance of ion balance.

R. P. Scheffer and colleagues have attempted to provide more direct evidence for the role of plasmalemma. Protoplasts isolated from susceptible oats lysed very rapidly (1 min) in the presence of HW-toxin (Samaddar and Scheffer, 1968). More recently, Rancillac *et al.* (1976) also reported lysis of susceptible protoplasts, but only to the extent of 80% at 4 h. Samaddar and Scheffer (1971) later approached the problem by application of chemicals (bisulphite, semicarbazide, hydroxylamine) to counteract HV toxin. The protectants were presumed to react with components of the plasmamembrane. Protection was afforded at high concentrations (10^{-3} to 10^{-5} M) but only in the presence of toxin. Although attempts were made to eliminate direct interaction between HV toxin and reagents, they are only partially convincing. Gardner and Scheffer (1973) reported that both cycloheximide and several-SH group reagents provided protection from HV toxin. Since the cycloheximide effect was reversible, they suggested a potential target might be an -SH dependent membrane protein which turned over rapidly. In 1970, Scheffer and Samadder reported that if dormant seeds were treated with victorin for 1 h before storage for various periods, viability was lost permanently in susceptible, but not resistant, varieties.

Scheffer and Samaddar (1970) suggested that since dormant seeds were affected, active metabolism was not necessary for HV toxin to exert an effect. This implies, then, that susceptible tissues contain a preformed site or receptor. However, no measures of metabolism (i.e. respiration) were studied either during aqueous victorin treatment or subsequent attempts at germination. Nonetheless, taken together, these results suggested irreversible physical damage to the plasma membrane.

In contrast, Wheeler and associates believed the site of action of

Victorin might be the region of the cell wall. Luke *et al.* (1966) could not find clear evidence from electronmicroscopy that plasma membranes were affected by victorin; although the membrane was swollen somewhat, the most pronounced change was the formation of inclusions in close juxtaposition to the wall. In another study, Hanchey *et al.* (1968) noted significant difference in reaction with uranyl ions between the walls of toxin-treated and control tissue. Hanchey and Wheeler (1969) also reported that HV toxin caused plasmolysis of treated cells in hypotonic solutions or in distilled water (false plasmolysis). It is difficult to reconcile this finding with protoplast lysis (Samaddar and Scheffer, 1968). Recently, McAmmon and Wheeler (1978) found that treatment of seed from susceptible varieties for 1 h followed by 1 h of extensive washing reduced germination from 99 to 20%, in agreement with earlier results of Samaddar and Scheffer. However, prolonged washing (16 h) or exposure of the seed to 0·16 M NaOH for 5 min (to inactive toxin) restored viability. Most significantly, alkali treatment was effective even after 2–3 weeks storage of victorin-treated seed. Similar results can be obtained with the less objectionable procedure of brief treatment with charcoal (McAmmon 1978).

Novacky and Hanchey (1974) and Gardner *et al.* (1974) observed rapid depolarization of plasma membranes in victorin-treated susceptible, but not resistant, cells. Novacky and Hanchey noted that depolarization was irreversible and thus physical bursting could not have occurred. In a more detailed study, Keck and Hodges (1973) reported that compartmental analysis indicated that properties both of plasmalemma and of tonoplast were affected at about the same time after addition of toxin. Again, their results suggest that bursting does not occur.

In all, the electrochemical evidence is no more definitive than any of the previous evidence attempting to establish the plasmalemma as a site for HV toxin action. To the authors, the uncertain state of the subject some 33 years after HV toxin was discovered has several origins. Failure to purify and chemically characterize HV-toxin has two major consequences. First, knowledge about structure might offer clues as to its mode of action as was the case for the non-specific toxin, rhizobitoxin (Giovanelli *et al.* 1971). Secondly, because purity cannot be established for any given preparation, it is not possible to determine how much of the effects are due to a combination of toxin with impurities. Pringle and Braun (1957) reported that 10 ng ml^{-1} completely inhibited root growth of susceptible plants. It is not possible to know whether the very high concentration of 16 μg ml^{-1} (Gardner and Scheffer, 1973) was used because of very low victorin content or because impurities play a role in the response. Similarly, the results of Saftner and Evans (1974) and Saftner *et al.* (1976) can not be evaluated satisfactorily without knowledge of purity (Daly, 1981). Expenditure of manpower and efforts on mode of action studies with impure preparations should be diverted in the future to attempts at purification and chemical characterization.

Evidence of a similar sort has been presented for a plasmalemma site for PC toxin affecting sorghum (Scheffer and Samaddar, 1970; Gardner *et al.*, 1972). Although O_2 consumption of tissue was increased by toxin, isolated mitochondria were not affected. Uptake of organic compounds was repressed and ion-leakage accelerated. Ion loss appears specific for K^+, since Na^+ loss was similar for resistant and susceptible tissues. As with HV toxin, cycloheximide appeared to protect susceptible sorghum tissues (Gardner *et al.* 1972), and the plasma membrane was depolarized (Gardner *et al.*, 1974). Nishimura *et al.* (1979) summarize similar evidence for an action of AK and AM toxins on plasma membranes. The direct demonstration of K^+ efflux should be noted.

Although it has been suggested that HC toxin affects the plasmalemma in susceptible corn, the results differ markedly from those from most other host-specific toxins. Yoder and Scheffer (1973a) reported increased uptake of NO_3^-, Na^+, Cl^-, 3-0-methyl glucose and leucine. Yoder and Scheffer (1973a) found that NO_3^- stimulated nitrate reductase activity but NO_3^- uptake was not dependent on synthesis of the enzyme. The fact that uptake is selective (K^+, Ca^{++}, PO_4^{-3} not affected) suggests that general increases in metabolism are not involved. Unfortunately this interesting system has not been studied in more detail.

As described in a later section, T toxin is known to interfere with oxidation of mitochondria from susceptible Texas male sterile corn (Miller and Koeppe, 1971). Arntzen and his colleagues, however, have dissented with the general view that the mitochondrion is the primary site of cellular damage. They argue for a primary effect on plasmalemma from the following considerations (Arntzen *et al.*, 1973): (a) inhibition of root growth occurred in 20 min, but respiratory increases required 2 h; (b) toxin caused a moderate decline in ATP and only after inhibition of root growth had occurred; (c) ^{32}P uptake or leakage was not significantly affected in susceptible roots by toxin, but general leakage occurred within 2 h. The protocol for these experiments is not without uncertainties (Daly, 1981), but the results suggest that inhibition of energy generation by mitochondria is not of primary importance in toxin action. It is essential to add the phrase: "at least in roots".

Work by Mertz and Arntzen (1972, 1977, 1978) on plasma membrane potential showed a rapid depolarization of root cells and selective inhibition of K^+, Na^+, Cl^- and PO_4^{-3} uptake caused by race T toxin. On the other hand, Keck and Hodges (1973) found no selective effects of toxin in efflux of Rb^+ from susceptible or resistant corn roots. Frick *et al.* (1976) found inhibition of K^+ uptake in midzone root sections, but only if toxin was present during an augmentation period of 90 min of aeration.

Interpretation of the above results is complicated by differences in procedures of each research group and by the unknown purity of the toxin preparations (Payne *et al.*, 1980a, Daly, 1981). With purified preparations, Payne *et al.* (1980a) found a rapid (10 min) increase in ion loss when root segments were aerated in the absence of toxin in order to reduce the endogenous efflux of ions to near zero. The toxin concentrations required were of the same magnitude as those which affected isolated mitochondria (Payne *et al.*, 1980a, b).

2. Mitochondria

Miller and Koeppe (1971) described host-specific effects of T toxin on mitochondria isolated from susceptible etiolated corn shoots, a significant landmark in studies of host–parasite interactions. The finding should lead directly to an explanation of toxin action in molecular terms, since the biochemical properties of mitochondria are reasonably well understood. Secondly, the identification of a precise cellular site logically should enable a clear statement about the biochemical role of T toxin in pathogenesis and especially in terms of damage from infection. So far these expectations have not been fulfilled.

The earlier descriptions of the effects of T toxin on mitochondria are reminiscent of a classic respiratory uncoupler, and the findings of Bednarski *et al.* (1977a, b) that ATP synthesis is inhibited *in vivo* and stimulated *in vitro* support this notion. Tipton *et al.* (1973, 1975) also reported inhibition of a microsomal K^+ activated ATPase, but unfortunately these observations have not been reaffirmed (Carl Tipton, personal communication). There are puzzling aspects of the action on mitochondria. Stimulation of NADH oxidation in the absence of ADP always is noted (Gregory *et al.* 1977, 1978), but malate oxidation is inhibited (Gregory *et al.* 1977, 1978). This would suggest (Mathews *et al.* 1978; Gregory *et al.* 1977) interference at or near the site of cytochrome B oxidation. Mathews *et al.* (1979) noted leakage of NAD^+ in toxin treated mitochondria and suggested that toxin disturbed the permeability of the inner membrane. As discussed in Section IV, A, 3, Watrud *et al.* (1975) reported that removal of the outer mitochondria membrane converted insensitive mitochondria to sensitivity. However, lipid analysis does not show significant difference between N and T mitochondria, although there appears to be a differential response to digitonin (Gregory *et al.* 1977a, b).

It is difficult to understand how leakage of essential inner membrane components could lead both to stimulation (on NADH substrate) and to inhibition (malate substrate) of oxidation. Further, succinate oxidation can be affected either way depending on whether the osmoticum contains high KCl (inhibition of oxidation) or low KCl (stimulation) (Gregory *et al.* 1977a). In our laboratory, the use of high KCl osmotica has not been satisfactory because the mitochondria are not as well coupled as in low KCl media (Payne *et al.* 1980b).

Even with low KCl media, additional complications arise. Although NADH oxidation appears uncoupled because mitochondria with stimulated O_2 uptake do not respond (or respond poorly) to added ADP, the same or slightly higher concentrations of toxin cause inhibition of malate oxidation, but O_2 uptakes appears coupled to ADP utilization (Payne *et al.*, 1980b). Interestingly, 500–1000 times greater concentration of T toxin causes inhibition of malate oxidation by mitochondria from resistant (N) corn or from soybean, but these concentrations, or higher, have no effect on succinate or NADH oxidation. It is possible to inhibit malate oxidation completely in N corn mitochondria and yet demonstrate coupled oxidation of succinate or NADH in the same mitochondria (Payne *et al.* 1980b). Because the toxin concentrations required to inhibit malate oxidation are much greater for resistant mitochondria, it might be argued that the effect on mitochondria involves a different site than that in susceptible Tms mitochondria. An acetate derivative (Kono and Daly, 1979) causes malate inhibition at the same concentrations in both N and Tms corn mitochondria, thus making this possibility unlikely (Payne *et al.* 1980b). The toxin acetate stimulates O_2 uptake with NADH and succinate in Tms corn mitochondria but, like native toxin, has no effect on the same process in N corn mitochondria (Payne *et al.* 1980b).

An unexpected finding by Payne *et al.* (1980b) was a loss in sensitivity of Tms corn mitochondria to toxin within 10–20 min after suspension of pellets in assay solutions. Pham and Gregory (1980) have observed a similar phenomenon, but their observations were made at 3, 6 and 9 h. Whether the same processes are involved remains an open question. The findings have an important bearing on the biology of host-specific toxins. If resistant and susceptible mitochondria differ in the presence or absence of a receptor site protein for example (Forde *et al.* 1977), it must be unstable, perhaps undergoing a conformational change with time. This would have important implications for any attempt to isolate and identify such protein. Secondly, if mitochondria can vary in sensitivity with experimental handling, it may be difficult to determine the degree of sensitivity of Tms mitochondria in cytoplasm *in vivo*.

The work of Belnarski *et al.* (1977b) with tissue indicates that mitochondria are sensitive *in vivo*, but there are some unexplained paradoxes. Although toxin increased respiration in most susceptible tissues studied, the respiration of root tips was unaffected. Payne *et al.* (1980a) did induce ion leakage in root tips. The uncoupler 2, 4-dinitrophenol acted, as expected, to stimulate glycolysis as well as respiration, but T toxin did not (Bednarski *et al.* 1977b). Although toxin caused a decline in cellular ATP, surprisingly an anaerobic period of 1 h was required to see the effect in treated susceptible tissues (Bednarski *et al.* 1977b). Walton *et al.* (1978) have provided the most direct evidence for a T toxin interaction with mitochondria *in vivo*. The ATP content of isolated mesophyll protoplasts in darkness drops quickly upon exposure to toxin but not when protoplasts are in light, presumably because photosynthetic ATP production masks the contribution of mitochondrial ATP synthesis (Earle *et al.* 1978).

3. Chloroplasts

Although bleaching of leaf tissue is a common end result of treatment with host-specific toxins, there are few data concerning the effects on photosynthesis or on the metabolism of isolated chloroplasts. Arntzen *et al.* (1973) reported a decline in CO_2 uptake in susceptible corn leaves within minutes after treatment with T toxin, but ascribed it to a secondary effect caused by stomatal closure. Bhullar *et al.* (1975) found an inhibition of both dark and light CO_2 fixation, but the long time (hours) required, plus distinct differences with the results caused by compounds known to close stomata, suggested a more direct effect on cellular processes. Interference with mitochondrial oxidation could cause inhibition of both types of CO_2 fixation, but the observation that portions of leaves banded by aluminium foil do not bleach with toxin treatment (Bhullar *et al.*, 1975) raises the question of a more direct effect on chloroplasts (Daly, 1981). Recently, Daly and Barna (1980) have shown rapid inhibition of dark and photosynthetic CO_2 fixation by T toxin in the thin leaf slices in which the contribution of stomatal closure should be minimal. Although the concentrations of T toxin required to see inhibition are in the same range as for effects on isolated mitochondria, there are important differences in the effects of toxin on light and dark CO_2 fixation. Inhibition of photosynthesis is apparent only upon 20–30 min of preincubation in light with toxin, while dark CO_2 is found in 10–20 min either with light or dark preincubation. This result is consistent with effects of light on necrosis (Bhullar *et al.* 1975).

Secondly, incubation under N_2 strongly reduces photosynthetic inhibition but has little effect on inhibition of dark CO_2 fixation. In combination, these results imply separate sites of action which may or may not involve disabling of mitochondria.

4. Interaction with Specific Host Compounds

Strobel (1973) reported that HS toxin was capable of binding to a protein fraction obtained from susceptible but not resistant sugar cane leaves. The basic evidence consisted of: (a) applying radioactive toxin preparations to leaf tissue and obtaining a small amount of radioactivity at or near the void volume of exclusion columns (Strobel, 1973); (b) preparation of protein or membrane fractions which appeared to bind radioactivity in equilibrium dialysis experiments (Strobel, 1973); (c) protection of susceptible tissues by treatment with galactosides which were presumed to be competitive analogues of HS toxin (see Section III, D); (d) transfer of the protein to tobacco cells which then were made susceptible to HS (Strobel and Hapner, 1975); (e) inhibition by HS toxin of transport of galactosides into isolated susceptible cells (Strobel, 1975).

This evidence suggested to Strobel (1975) that the protein was a component of a transport system required for galactose-containing compounds. It is difficult to understand, however, why the system would be important in disease. Although resistant cells presumably do not contain such a protein (Strobel, 1973), its absence obviously is not detrimental to the function of healthy resistant plants. Other experiments (Strobel, 1976) suggested it might have a dual function as an ATPase, but the specific activities reported for the enzyme are two to three orders of magnitude less than other purified plant ATPase (Leonard and Hodges, 1973).

There has been no independent confirmation of any of these claims and, unfortunately, no new, more convincing evidence from any source. Wheeler (1976) and Daly (1976) have pointed out some uncertainties in the techniques employed. Since the existing evidence recently has been examined in some detail (Daly, 1981), it seems unnecessary to furnish more than an outline of the principal uncertainties. (a) The bioassay employed appears to be quite variable and insensitive. The data (Strobel, 1973) purporting to show protection by galactoside are based on control values showing large innate differences (Bhullar *et al.*, 1975). (b) Most of the crucial experiments used toxin preparations of very low specific radioactivity (Daly, 1976, 1981). This permits questions about the validity of the *in vivo* binding demonstration for susceptible tissues as well as the accuracy of *in vitro* determination of binding constants (Daly,

1981). (c) Even if the level of radioactivity employed was experimentally permissible, there are significant errors in the calculation of the binding constants (Daly, 1976). (d) The given values of $1{\cdot}1 \times 10^{-5}$ M for the dissociation are in the same order as for enzyme-subsubstrate binding, and are several orders of magnitude lower than the values reported for biologically potent compounds such as hormones (10^{-10}, Gorski *et al.* 1968), Tentoxin ($3{\cdot}2 \times 10^{-9}$ M, Steele *et al.* 1976) and Fusicoccin (10^{-8} M, Dohrmann *et al.* 1977). It seems unlikely that binding constants as low as those reported for HS toxin would be useful as tracers for isolation of a very specific protein from leaves. If valid, the binding constants may explain the very low innate toxicity of these preparations. Although the data were expressed in terms of nanomoles, the solutions applied to the bioassay curve (Steiner and Strobel, 1971) are very concentrated.

In a recent abstract, McFarland and Gilchrist (1980) have postulated that host-specific toxin from *A. alternata* sp. *lycopersici* may inhibit aspartate carbamyl transferase (ACTase). Protection from the toxin was afforded to tissues treated with members of the aspartate metabolic family, including orotic acid. ACTase activity in extracts were selectively inhibited. A significant finding (McFarland and Gilchrist, 1980) is that this host-specific toxin does not cause rapid ion leakage from tissues. Rather, leakage occurs after several days with the onset of visual symptoms. This result differs markedly from other specific toxins and may indicate a different mode of action.

V. CONSIDERATIONS FOR FUTURE RESEARCH

The foregoing summary indicates clearly that fundamental knowledge about host-specific toxins at the molecular level is limited. This undoubtedly is a reflection of the relative newness of this area of research and also, perhaps, a high innate degree of experimental difficulty. Certainly the marked specificity of these toxins also is characteristic of other host–pathogen interactions which do not involve host-specific toxins. No completely satisfactory chemical explanation exists in these cases either, despite a considerably longer history of investigation. It also is possible, however, that progress has been slower than anticipated because early results have channelled research in directions which should be re-examined. Reflections on the past literature suggest some future emphasis along different lines. Certain procedural possibilities were discussed in Section IV, A. It seems appropriate in the following section to offer speculative views on the conceptual issues that so far are unresolved.

A. On Modes of Action

1. Similarities Among Host-Specific Toxins

Certain similarities in the action of host-specific toxins on susceptible hosts implicitly are acknowledged in the literature, but the similarities have not been exploited experimentally. Data outlined in Section IV on the four most extensively investigated toxins (HV, PC, HC, and T toxins) shows that three (HV, PC and T toxins) stimulate respiration, inhibit growth, cause ion leakage (especially K^+ flux) and depolarize plasma membrane potential differences (PD). In contrast, HC toxin stimulates growth and promotes Na^+ and NO_3^- uptake (Yoder and Scheffer, 1973a, b). In addition, T toxin (Arntzen *et al.*, 1973) and HV toxin (Turner, 1972) cause a rapid, but apparently transient, stomatal opening. There are little or no data on the direct effects of HC and PC toxins on stomata.

The non-specific toxin Fusicoccin (Fa) (Ballio *et al.*, 1964) exhibits a notable antiparallelism to the action of HV, PC and T toxins: growth stimulation (Ballio, 1978), promotion of K^+ and Na^+ uptake (Cocucci *et al.*, 1977), hyperpolarization of PD (Cocucci *et al.*, 1976), opening of stomata (Turner and Graniti, 1969). Where comparisons can be made, HC toxin, on the other hand, parallels FC toxin. In addition, both FC toxin (Marre, 1979) and HC toxin (Kuo and Scheffer, 1970) stimulate dark CO_2 fixation, while T toxin inhibits dark CO_2 fixation (Bhullar *et al.*, 1975; Daly and Barna, 1980). Although there is a single brief report of stimulation of dark CO_2 fixation by HV toxin (Kuo and Scheffer, 1967), we have found only inhibition of fixation in susceptible, but not resistant, oat varieties in our laboratory with procedures similar to those employed for T toxin (Daly and Barna, 1980). Such an effect is more consistent with the opposing effects of HV and HC toxins in other processes. In the case of HV toxin, the antiparallel behaviour of FC toxin can be extended to changes in the C_6/C_1 ratio. FC toxin stimulates the ratio (Marre, 1979), HV toxin decreases it in susceptible tissues (Rawn, 1977).

2. The Role of Energy-Transducing Membranes

FC toxin represents one of the best understood plant pathotoxins in molecular terms. Marre (1979) has recently summarized an extensive literature. The available evidence suggests that FC toxin binds (Dohrmann *et al.*, 1977) at, or near, a K^+-activated plasmalemma ATPase (Beffagna *et al.*, 1977), causing proton extrusion (Lado *et al.*, 1975, 1976) accompanied by counter influx of K^+ ions (Lado *et al.*, 1976) and hyperpolarization of the plasmalemma. Proton extrusion

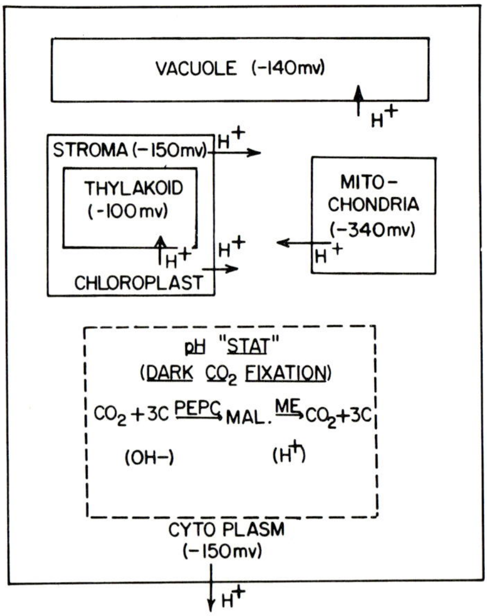

Fig. 6 Idealized membrane potentials and direction of proton flux in a photosynthetic cell.

leads to acid-induced fast growth (Marre, 1979) while change in K^+ flux affects stomatal opening (Fisher, 1968; Rasche, 1975). Stimulation of CO_2 fixation by FC toxin may result from proton extrusion, raising cytoplasmic pH to activate PEP carboxylase in its role of servicing a cytoplasmic "pH stat" (Davies, 1979; Raven and Smith, 1974). The formation of malate by dark CO_2 fixation could lead to higher C_6/C_1 ratios by altering the ratio of regulatory NADPH and NADP (Marre, 1979). Thus, a single event at the plasmalemma can lead to extensive metabolic ramifications. These postulated relationships are diagrammed in detail by Marre (1979).

Figure 6 provides similar considerations but with emphasis on plant membranes in terms of the flux of protons during energy transduction. The usually obligatory accompanying, or countering, flux of other ions is ignored because specific details are not known in most cases.

The antiparallel affect of HV, PC and T toxins, when compared with FC toxin might indicate fundamentally the same basic targets on the plasmalemma, but with changes in the opposite direction for the increased flux of protons and inorganic or organic ions. Alternatively, the initial effects may be directionally quite similar, but the required cytoplasmic compensation for the outward flow of protons may not occur, leading for example, to growth inhibition rather than growth stimulation. Such an alternative may explain the parallel behaviour of HC and FC toxins in short term experiments. The basic initial action of both may be very similar, but the long-term consequences are, of course, quite different. The observations of Bronson and Scheffer (1977) and Nishimura *et al.* (1979) on decreased sensitivity to host-specific toxins by brief treatments with heat might be interpreted as a result of conformational changes in membranes.

The fact that T toxin affects isolated mitochondria has focussed attention on this organelle, rather than on plasmalemma or other membranes, as a primary site of action, although evidence to the contrary is cited in Section IV.B.1. A more intriguing possibility is that T toxin, as well as other host-specific toxins, possesses the general property of affecting all cellular energy-transducing membranes (Daly, 1981; Daly and Barna, 1980; Payne *et al.*, 1980b). Even a primary site on mitochondria can be reconciled because the primary changes in ion flux in this organelle will affect (a) fluxes at the plasmalemma through cytoplasmic changes in ion concentrations (Fig. 5) or (b) production of mitochondrial ATP necessary for plasma membrane functions. The observation that sensitivity of isolated mitochondria to T toxin markedly decays in short times (20 min) opens the possibility that HV toxin, for example, does affect mitochondria but that the isolation procedures may have prevented observations of such effects (Wheeler and Hanchey, 1966).

Regardless of the selectivity for a particular membrane, it seems obvious from Fig. 6 that the interference with any energy-transducing membrane can be lethal to sensitive tissue and will result in a commonality of physiological symptoms among even highly specific toxins. Further, such a general mechanism may account for the observation (largely uncommented upon previously) that, although such toxins are highly specific for certain varieties, all organs and tissues of a sensitive variety are affected by them. If a specific protein, for example, is the target for a host-specific toxin, the available evidence suggests that it is present in all cells and tissues regardless of their function. Raven and Smith (1976) have noted that membrane function in ion transport varies quantitatively between specialized root and leaf cells by virtue of the external environment, but that the basic molecular activities are essentially the same.

3. Ionophorous Antibiotics

Some antibiotics are known to function as ion or proton translocators of membranes (Pressman, 1976). Knowledge of the chemistry and action of ionophorous antibiotics has progressed markedly in the last decade, following the initial work of Pressman and his colleagues on valinomycin (Pressman, 1963; Harris and Pressman, 1969). In all cases they appear relatively non-specific for the membrane affected, but do show specificity in the nature of the ion transported or translocated through a membrane. In general, the structures of ionphorous antibiotics must permit either sequestering of an individual ion or else the formation of holes, or channels, in the membrane itself. Because of the paucity of information on structures of host-specific toxins, it is not possible to apply such concepts directly to them. Kono and Daly (1979) and Payne *et al.* (1980b) have noted that the structures of T toxin (Fig. 2) suggest that they could function as ionophores. Their lengths approximate to that of a membrane bilayer and they can assume configurations with separable hydrophilic (oxygen functions) and hydrophobic centres. Lack of precise information on structures does not preclude, however, investigation of these possibilities, since much of the information on ionophores was obtained by studies on ion translocation in natural or synthetic lipid bilayers (Pressman, 1976).

4. Toxins and Hormones

The similarities in the action of FC toxin and IAA have led to the suggestion that they may affect the same or a very similar cellular or membrane site (Marre, 1979), but binding experiments with isolated plasma membranes show no competition between them. Recently, Federico *et al.* (1980) have reported that T toxin antagonizes the effects FC toxin in susceptible, but not resistant, corn roots and thus may act near, or on, the same site as FC and IAA. In an earlier study of HV toxin Saftner and Evans (1974) made a similar suggestion. They reported a specific but transient growth stimulation with HV toxin on susceptible oats.

Potential relationships between the action of toxin and the action of hormones deserves more attention in the future. There is a growing awareness that of the action of endogenous plant hormones may lie at critical centres in energy-transducing membranes. Obviously, interference by toxins (non-specific as well as specific) will have a disproportionate effect on cellular metabolism because of the catenary or cascading nature of hormonal control. Certainly, interference with basic hormonal control could be a lethal device. In addition, the very low concentration of natural hormones required for activity suggests a limited number of required cellular targets. Host-specific toxins of high purity exhibit effectiveness at extremely low concentrations, usually less than 10 ng ml^{-1} (Livingstone and Scheffer, 1980, Kono and Daly, 1979; Nishimura *et al.*, 1979; Payne *et al.*, 1980b; Dunkle and Wolpert, 1981).

There are at least two molecular analogies in animal systems. Cholera and diphtheria toxins both appear to affect membrane bound AMP-cyclase in much the same way as hormones such as epinephrine (Moss and Vaughn, 1979). The potential role of hormones in plant diseases has been discussed elsewhere (Daly, 1976b; Daly and Knoche, 1976).

B. On the Nature of Specificity

1. Recognition (Receptor) Sites

The early structural work on HV, PC and HC toxins suggested that these contained peptide bonds. Since the biological specificity inherent, in polypeptide structures is implicit in all concepts of biochemistry, these early studies implanted the notion that the toxin molecule made a significant contribution to the observed specificity. Because only AM toxins have been shown beyond doubt to possess peptide bonds, the validity of this idea is open to question. The basic structure of T toxin does not readily suggest how specificity in its action is achieved. A second aspect of earlier research was the evidence that resistant tissues were not affected significantly by host-specific toxins. Thirdly, certain evidence (Samaddar and Scheffer, 1968, 1970; and Scheffer and Samaddar, 1970) suggested that host metabolic activity was not required for toxin sensitivity. This evidence indicated that susceptibility was conditioned by the presence of a host component which, because the toxin acted in the absence of the pathogen, was not synthesized nor activated during the infection process (Scheffer, 1976).

These lines of evidence thus lead to the logical conclusion that specificity is dependent on interaction (binding) of toxin with a unique host component present only in susceptible tissues. The main alternative is that insensitivity results from active toxin degradation in resistant, but not susceptible, tissue (Wheeler, 1975). This idea was tested (Romanko, 1959), but the positive results obtained were not confirmed (Scheffer and Pringle, 1964). Wheeler's (1968) attempt to resolve this issue was inconclusive because of the necessity to use toxin preparations of unknown purity and without the availability of a quantitative chemical test for degradation.

Recent work by Payne *et al.* (1980b) with T toxin on resistant corn mitochondria (see Section IV. B. 2.) have a bearing on this problem, but also do not provide a completely satisfactory answer. Since inhibition of malate oxidation occurs in resistant mitochondria, T toxin must be penetrating to and affecting the same site as in susceptible mitochondria. The higher concentration required might be explained by (a) a different affinity constant for very similar sites in both types of mitochondria or (b) some degradation or alteration of toxin in resistant mitochondria that reduces its effectiveness. For the latter explanation, it is difficult to envisage the nature of the processes concerned, especially if host enzymes are involved. It is interesting to note that *P. maydis* toxin also inhibits malate oxidation of resistant mitochondria, an observation which originally was attributed to impurities (Comstock *et al.*, 1973). There is no biological or chemical evidence for impurities in T toxin used in the work of Payne *et al.* (1980a, b).

The most direct evidence for the presence of specific receptor sites would be to demonstrate binding to organelles or isolated cellular components, most readily accomplished with radioactive toxins. Lack of assurance about purity prevent such studies for most host-specific toxins. Even for known structures, a problem remains. Because of their potency (1–10 ng ml^{-1}), a high level of radioactive labelling is required and the cost may be prohibitive. With T toxin, for example, the minimum amount of labelling required for an experiment to establish whether or not binding occurred would be 1000 dpm ng^{-1}, and 10 000 dpm ng^{-1} probably would ensure an adequate test. It is possible to synthesize, by routes devised by Y. Suzuki (Section III. G), species of the structures shown in Fig. 3, especially compound IV which is active, like native toxin, at 10 ng ml^{-1}. We were disappointed to calculate that synthesis of 100 mg of compound IV would require $500 000 for radioactive precursors, even if the level of radioactivity was only 1000 dpm ng^{-1}.

Another but less satisfactory approach for identifying a specific recognition site would be the demonstration, with techniques such as gel electrophoresis, of unique macromolecules in susceptible, but not resistant, tissue. Ford *et al.* (1978) reported that mitochondria from Tms corn contained 21 polypeptides while resistant N corn mitochondria possessed only 20. However, the additional component was postulated only because an additional ^{35}S-labelled component could be detected. Protein staining, which showed the 20 polypeptides present in both types of mitochondria, did not reveal any additional bands in Tms corn. The significance of the observation thus is open to question.

2. *Wheeler's Self-Repair Hypothesis*

The receptor site hypothesis to explain host-specific effects was developed primarily from early experience with victorin, which suggested that this toxin did not affect resistant hosts in any significant way. Wheeler and Doupnik (1969) found that high concentrations of victorin did induce physiologic changes in resistant hosts, but did not cause visible symptoms. Since the physiologic changes were similar to those in susceptible hosts, these observations and others led Wheeler (1969, 1975) to suggest that toxin initially might effect a similar site in both resistant susceptible cells, but that resistant hosts underwent self-repair or recovery.

This important notion generally has been overlooked, but should be considered more carefully in the future. It implies that cell recognition of the toxin is not necessary. Rather, the initial biochemical events may be identical. Specificity lies in how the cells adjust biochemically to the initial event. As far as we are aware, there is no experimental evidence to rule out this possibility, and it can be accommodated easily in the discussion of Section V. 5. However, convincing evidence in its support can come only from experiments with toxins of assured purity. It should be noted also that crucial earlier data lending support to the receptor hypothesis has been questioned in recent years. The work of McAmmon and Wheeler on HV toxin is cited in Section IV. B. 1., for example. Although the self-repair hypothesis eventually may be found to apply only to certain host-specific toxins, it should influence the direction of research for all of them.

3. *Genetic Aspects*

The genetic control of disease reaction has a bearing on the interpretations of both the mode of action and the nature of the specificity of host-specific toxins. The molecular events eventually have to reconcile with the mode of inheritance (Daly, 1972). Flor's (1955) hypothesis that single complementary, dominant alleles in both host and parasite are required for the expression of resistance of flax varieties to races of *Melampsora lini* has generally been interpreted to mean that an active gene product (for resistance) in the host interacts with an active gene product (for avirulence) of the pathogen. Further, it is argued that this combination alone is responsible for the specificity of disease reaction since it is the only unique quadrant in the four possible combinations of genotypes from homozygotes of host and pathogen (Ellingboe, 1976, 1979). Flor's hypothesis has been explored primarily with diseases caused by biotrophic pathogens, but whether the above molecular interpretation is valid even for these diseases has been questioned (Daly, 1972, 1976). Wheeler (1969, 1975) has raised the philosophical point that any interpretation of genetic data is conditioned by one's assumptions; particularly, in this case, by assumptions about the meaning of dominant and recessive inheritance.

In any event, the biology of host-specific toxins appears at variance with this molecular interpretation. At first glance, host-specific toxins appear to be active gene products for virulence, not for avirulence. Secondly, in two well documented cases (HV and PC toxins), resistance of the host to both disease and to toxin is conferred by a recessive gene (Scheffer, 1976). Resistance to black spot of pear (AK toxin) and the Alternaria blotch of apple (AM toxin) also is inherited as a recessive trait (Nishimura *et al.*, 1974). Resistance to HC toxin and to the disease is controlled by a major dominant gene with two alleles which provide intermediate resistance. Minor genes further modify resistance, but in all genotypes tested (Kuo *et al.*, 1970), the action of toxin parallels disease reaction. Although genetic data for the most recent cases of host-specific toxins (Table I) are not available, none of the best-known examples appear to fit precisely with Flor's postulates.

Applying Wheeler's logic, however, this outcome is dependent upon how modes of action are interpreted. Litzenberger (1949) equated sensitivity to victorin as an extreme case of hypersensitivity with the implication that the dominant allele for disease reaction is, in fact, producing a product causing extreme resistance. Flor's postulates thus would apply. Wheeler (1969) presents arguments against this concept. Litzenberger was influenced by the knowledge that the allele for susceptibility to *H. victoriae* is either the same allele as the dominant allele controlling resistance to certain crown rust races of a separate closely linked gene (Litzenberger, 1949). As far as we are aware, this intriguing association has not been examined physiologically or biochemically. It would be interesting to know, for example, what influence various concentrations of HV toxin would have on the expression of a disease in varieties resistant or susceptible to crown rust fungi.

If Wheeler's self-repair hypothesis is correct, Flor's postulates also may apply. Self-repair might involve different and additional repair enzymes in resistant, but not susceptible, cells. However, it is not necessary to invoke additional proteins for self-repair. Controlling processes for proton or ion flux (Section V. A.) employing enzymes common to both resistant and susceptible tissue can be invoked.

Because of the nature of the life cycles for the pathogens listed in Table I, studies of the genetics of toxin production are limited. The work of Scheffer *et al.*, (1967) is particularly illuminating. *H. victoriae* and *H. carbonum* are sexually compatible and crosses between toxin producers of each species resulted in four equal classes of phenotypes: production of HV toxin only; production of HC toxin only; production of both toxins by single strain; neither toxin was produced. The findings indicate separate single gene control for production of each toxin. The fact that both genes can exist in an individual strain has an important bearing on disease development in nature. Nishimura *et al.* (1979) have provided a very interesting account of the potential interrelationships between toxin production and host range in *Alternaria* species.

There are a number of interesting physiological problems that might be explored if the appropriate genetic materials become available. In section V. A. 1, we noted the contrasting effects of T and HC toxins on processes in corn, such as ion leakage and dark CO_2 fixation. Inclusion of the allele for susceptibility of *H. carbonum* into lines containing Tms cytoplasmic factors for susceptibility to *H. maydis*, race T would provide uniform genetic material for testing the effects of each toxin alone, but particularly in combination, on such processes. It might be rewarding if the genes for toxin production in the Helminthosporia could be aligned in various combinations in individual strains. This might be advantageous in studies of biosynthesis as well as for practical problems in toxin production and isolation.

Attention also should be directed to the interesting observation of Odvody *et al.* (1977). Roots of resistant and susceptible sorghum cultivars were obtained from a disease nursery in Kansas known to be heavily infested with *P. circinata*, and a number of pathogen isolates were obtained. Thirty four percent of approximately 100 isolates from susceptible plants were toxin producers, while none of the 200 isolates from resistant plants produced toxin, even after several generations of subculture. These results subsequently have been confirmed with isolates obtained from a different disease nursery in Texas. They raise important questions about the biology of the *P. circinata*-sorghum interactions in nature and especially about the role of host-specific toxins in the infection process. There is a clear correlation between susceptibility to disease and sensitivity to toxin. Why resistant plants yield only non-producers is unknown, but one intriguing possibility is that resistant plants have the ability to turn-off toxin production during infection in addition to being insensitive to toxin. If so, it will be difficult to explain single gene control of disease reaction (Schertz and Tai, 1969) because it is logical to assume that a turn-off mechanism would be under different genetic control.

C. On Roles in Pathogenesis and Parasitism

The experimental protocols used in demonstrating the existence of host-specific toxins by itself constitutes reasonable evidence for a role in plant pathogenesis, but such evidence is at best qualitative. The quantitative importance of a host-specific toxin for the spectrum of events during disease progression is not easily evaluated. As noted in other sections, the early work with victorin may have disproportionately influenced ideas concerning the role of host-specific toxins in nature. Victorin reproduced all of the known physiologic and visual symptoms found in naturally infected tissue. It also was produced by germinating spores (Nishimura and Scheffer, 1965) suggesting an action at the time of penetration of the host by *H. victoriae*. AK toxin (Nishimura *et al.*, 1979) and AM toxin (Kohomoto *et al.*, 1976) and also appear to be present in germinating spores.

Interpretation of the work of Yoder and Scheffer (1969) is especially significant in evaluating the role of victorin in disease. When excised leaves of susceptible oats were inoculated with a non-producing strain of *H. victoriae* or with *H. carbonum*, a non-pathogen, the histological events resembled those in oats resistant to *H. victoriae*. Penetration occurred but growth of the pathogens was limited to only several cells. When victorin was supplied, development of both avirulent *H. victoriae* and *H. carbonum* was like that of toxin-producing virulent strains on susceptible oats, suggesting that parasitism itself was dependent on the toxin.

In evaluating these data, however, it should be remembered that histological observations are difficult to analyse quantitatively, and the results may be dependent on the sites selected for inoculation or observation. A chemical estimation of fungal development would be more reassuring (Daly, 1976). Secondly, in our experience, inoculation of susceptible leaves did not result in significant lesion or symptom development, even with virulent, toxin-producing strains. Although no visual symptoms were seen on resistant varieties, only a few very small, white lesions (less than 0·5 mm) were found on susceptible leaves. In nature, *H. victoriae* is a root and stem pathogen, although producing extensive foliar damage. The use of excised leaf tissue therefore may not be appropriate for evaluating the role of victorin in either pathogenesis or parasitism. Obviously, failure to see typical symptoms when leaf tissue is inoculated implies that factors other than toxin are involved during pathogenesis in root or stem infections. Additional pathogen or host compounds may be translocated into shoot tissues. It is possible, of course, that toxin production is turned off in leaves (see Section V. B. 3).

There is one other report suggesting that growth and development of a pathogen is related to the effects of host-specific toxins on host cells. Lim and Hooker (1972) reported that sporulation of *H. maydis* race T occurred much earlier and was abundant on susceptible, but not resistant, corn.

There can be little argument that host-specific toxins account for most, if not all, of the cellular damage that occurs during infection of susceptible tissues. This is evident in the runner symptoms of Southern corn leaf blight and eye-spot, or frogs-eye, disease of sugar cane. In these diseases, the toxins appear to represent added factors for virulence, since significant lesions are formed on resistant host varieties. In fact, one of us (J.M.D.) has observed consistently larger primary lesions caused by *H. sacchari* on sugar cane clones classed as resistant than the primary lesions formed on susceptible clones. The runner symptoms were absent on resistant clones. With Southern corn blight, however, primary lesions generally are smaller on resistant N cytoplasm under our conditions of growth. In the case of the T-like toxin produced by *Phylosticta maydis*, it does not appear to be the determining factor in the expression of damage or disease caused by this pathogen on corn varieties. Its influence is observed when *P. maydis* infects tms corn, but there obviously are other factors controlling pathogenesis in this instance (Comstock *et al.*, 1973).

To resolve these important questions, more emphasis, especially quantitative emphasis, should be placed in on the biology and infections for diseases in which host-specific toxins are involved. There may be unsuspected aspects of parasitism and pathogenesis that such studies will reveal.

VI. ADDENDUM, SEPTEMBER 1981

Kenfield and Strobel (1981a) have reported very recently on the binding of raffinose, galactinol and pathotoxin from *Helminthosporium sacchari* (HS-toxin) in protein extracts from sugar-cane, mint and tobacco. As in previous work from this laboratory, the data in Tables I, III and IV of the paper were obtained by determining the amount of radioactivity remaining in a 0·5 ml half cell after equilibrium dialysis against 0·5 ml of protein preparation. No primary data were given; the final results are expressed as "specific binding activity" in μmol bound ligand/mg protein (Table I), mmol bound/g protein (Table III), or dissociation constant (K_D) values in Table II of Kenfield and Strobel (1981a).

Binding appeared to exhibit anomalous behaviour during purification and especially with apparently purified protein. For example, binding of raffinose to sugar cane protein was 370 times greater at 0·5 μg ml^{-1} than with 5 μg ml^{-1} of purified protein and 2230 times better than for crude preparations at 4 mg ml^{-1}. The latter value is in the range of protein concentrations used in early studies (Strobel, 1973) of HS-toxin binding to protein. The absolute masses involved can be determined readily from the mol wts of raffinose (504) and galactinol (344). The mol wt of HS-toxin was given as 584, based on an unpublished pH.D. thesis of Beir, but is at variance with the value of 236 used in previous work.

HS-toxin bound significantly to proteins from all three species. At

$2{\cdot}2 \times 10^{-5}$ M concentrations of ligand, raffinose and HS-toxin were bound to sugar cane protein to the extent of 111·60 and 35·60 μmol per mg protein at a concentration of 0·5 μg per ml (0·25 μg protein in 0·5 ml half cell). Nearly identical binding values are given for HS-toxin binding in Table III. Such binding is equivalent to 56·30 mg raffinose or 20·76 mg HS-toxin per mg protein. In a related paper, similar values for HS-toxin are given in Table I of Kenfield and Strobel (1981b) (15 μg toxin per μg protein).

The mol wt of binding protein from the three species was reported as slightly over 100 000 daltons. The first report (Strobel, 1973) on binding protein from sugar cane claimed a size of 49 999 daltons which was used in several later papers. It contained four subunits of ≈17,000 daltons. If the current value of 100 000 daltons is assumed, the molar binding ratios of raffinose and HS-toxin to sugar cane protein are approximately 11 100 and 3560, respectively. Assuming an average weight of 120 for amino acids in the protein, there were 13·3 molecules of raffinose or 4·27 molecules of HS-toxin bound per amino acid residue. This is extraordinarily high ligand binding. For HS-toxin, the results are in marked contrast to the total of 2–3 sites claimed for a binding protein of 49 000 daltons (Strobel, 1973).

Binding to the extent calculated above raises questions about the protocol used to determine the extent of binding. For example, crude protein at 4 mg mol^{-1} bound HS-toxin ($2{\cdot}2 \times 10^{-5}$ M or 12·8 μg per dialysis cell) to the extent of 1·09 μmol mg^{-1} protein. Since 2 mg of protein were present in the dialysis cell, the protein should bind a total of 1273 μg of HS-toxin, far in excess of the amount supplied. A similar uncertainty exists for raffinose (11·1 μg dialysis $cell^{-1}$) binding by 0·5 μg ml^{-1} protein. Since the amount of protein present is only 0·25 μg, binding of 111.60 μmol ligand mg^{-1} protein would require at least 14·1 μg. Because the dissociation constant is not particularly low ($K_D \approx 10^{-5}$ M), an appreciable fraction of free raffinose will remain in solution, in equilibrium with the 14·1 μg theoretically bound to protein.

The new specific binding activities for HS-toxin are more than 1000 times higher than the values reported earlier (Strobel, 1973), and it is surprising that the apparent K_D for protein from susceptible sugar cane (3×10^{-6} M) is only 3 to 4 times lower than in the original report ($1{\cdot}1 \times 10^{-5}$ M). Interpretation of the role of toxin in disease also is complicated by the fact protein from resistant sugar cane has a similar K_D of $1{\cdot}2 \times 10^{-5}$ M, although this sugar cane line (H50–7209) previously was reported not to bind HS-toxin (Strobel, 1973). Further, Table IV of Kenfield and Strobel shows that the specific binding activities of both resistant and susceptible cane are identical.

The calculations employed to estimate K_D values are not indicated, but the high values may result from the fact they were determined with protein at 3 to 4 μg ml^{-1}. In a separate paper, Kenfield and Strobel (1980) tested protein concentration from 0·01 to 1·0 μg per ml and report that raffinose binding is maximal to 0·05 μg ml^{-1} and drops sharply to 1/10 maximal activity at 0·4 μg ml^{-1} or greater concentrations. The same effect can be seen in Table I of their later (1981a) report.

It is possible that the high degree of apparent binding is the result of a systematic error in calculations used for the tables. For example, a statement in the text indicates binding of HS-toxin to two purified proteins was 2·2 and 2·0 μg toxin mg^{-1} protein, about 10 000 less than the values given in the tables and similar to values reported earlier (Strobel, 1973). However, problems arise in accepting the text values when the protocol is examined.

Only two μg of protein was assayed in the presence of 1 ml of radioactive HS-toxin at $2{\cdot}2 \times 10^{-5}$ M (specific activity 21·8 nCi μmol^{-1} or 82 dpm per μg HS toxin). Binding at the rate of 2·2 μg per mg protein would result in binding of 0·004 μg HS-toxin by the 2 μg protein present in the half cell, a total binding of only 0·36 dpm of radioactive label. Experimental detection of such an amount seems quite unlikely.

Although there are technical uncertainties concerning these results, if they are correct a number of problems arise in understanding the mechanism of action, and the role in disease, of toxin and/or protein binding. A major argument for a role in disease was the strong correlation reported for disease susceptibility and the extent of binding to protein (Strobel, 1973). Another line of evidence for a role in disease was the conversion of tobacco protoplasts from toxin resistance to susceptibility by incorporation of binding protein from susceptible sugar cane (Hapner and Strobel, 1976). Such earlier results are a variance with the new finding that both tobacco and resistant sugar cane contain similar amounts of a binding protein 10 000 times more effective than previously reported for protein from susceptible sugar cane. Further, its isolation from mint implies it may be ubiquitious in higher plants.

REFERENCES

Aranda, G., Berville, A., Cassini, R., Fetizon, M. and Poiret, B. (1978). *Ann. Phytopathol.* **10,** 375–379.

Arntzen, C. J. (1972). *Biochem. Biophys. Acta* **283,** 539–542.

Arntzen, C. J., Haugh, M. F. and Bobick, S. (1973). *Plant Physiol.* **52,** 569–574.

Arntzen, C. J., Koeppe, D. E., Miller, R. J. and Peverly, J. H. (1973). *Physiol. Plant Path.* **3,** 79–89.

Auriol, P., Strobel, G., Pio Beltran, J. and Gray, G. (1978). *Proc. Natl. Acad. Sci.* **75,** 4339–4343.

Ballio, A., (1979). *In* "Advances in Pesticide Science", (H. Geissbühler, ed.) Part II, pp. 336–372. Pergamon Press, Oxford.

Ballio, A., Chain, E. B., DeLeo, P., Erlander, B. F., Mauri, M. and Tonolo, A. (1964). *Nature,* London **230,** 230–297.

Bednarski, M. A., Izawa, S. and Scheffer, R. P. (1977a). *Plant Physiol.* **59,** 540–545.

Bednarski, M. A., Scheffer, R. P. and Izawa, S. (1977b). *Physiol. Plant Pathol.* **11,** 129–142.

Beffagna, N., Cocucci, S. and Marré, E. (1977). *Plant Sci. Lett.* **8,** 91–98.

Bennet, J. (1976). *Phytochemistry* **15,** 263–265.

Berville, A. (1978). *In* "Plant Mitochondria", (G. Ducet and C. Lance, eds) Elsevier, Amsterdam.

Bhullar, B. S., Daly, J. M. and Rehfeld, D. W. (1975). *Plant Physiol.* **56,** 1–7.

Black, H. S. and Wheeler, H. (1966). *Amer. J. of Bot.* **53,** 1108–1112.

Bottini, A., Bowen, J. R. and Gilchrist, D. G. (1981). *Tetr. Letts.* **22,** 2723–2726.

Bottini, A. and Gilchrist, D. G. (1981). *Tetr. Letts.* **22,** 2719–2722.

Bronson, C. R. and Scheffer, R. P. (1977). *Phytopathology* **67,** 1232–1238.

Cocucci, S. and Cocucci, M. (1977). *Plant Sci. Lett.* **10,** 85–95.

Cocucci, M., Marré, E., Ballarin Denti, A. and Scacchi, A. (1976). *Plant Sci. Lett.* **6,** 143–156.

Colchasure, G. E. and Yapp, J. H. (1976). *Physiol. Plant Pathol.* **37,** 298–302.

Comstock, J. C., Martinson, C. A. and Gengenback, B. G. (1973). *Phytopathology* **63,** 1357–1361.

Comstock, J. C. and Scheffer, R. P. (1973) *Phytopathology* **63,** 24–29.

Daly, J. M. (1972). *Phytopathology* **62,** 392–400.

Daly, J. M. (1976). *In* "Physiological Plant Pathology", (R. Heitefuss and P. H. Williams, eds) pp. 27–50. Springer-Verlag Berlin, Heidelberg and New York.

Daly, J. M. (1976). *In* "Specificity in Plant Diseases", (R. K. S. Wood and A. Graniti, eds) pp. 231–234. Plenum Publishing Corp., N.Y.

Daly, J. M. (1976). *In* "Specificity in Plant Diseases", (R. K. S. Wood and A. Graniti, eds) pp. 151–167. Plenum Publishing Corp., N.Y.

Daly, J. M. and Barna, B. (1980). *Plant Physiol.* **66,** 580–583.

Daly, J. M. and Knoche, H. W. (1976). *In* "Biochemical Aspects of Plant–Parasite Relationships", (J. Friend and D. R. Threlfall, eds) pp. 117–131. Academic Press, London and New York.

Davies, D. D. (1979). *Ann. Rev. Plant Physiol.* **30,** 131–158.

de Mayo, P., Spencer, E. Y. and White, R. W. (1961). *Can. J. Chem.* **39,** 1608–1612.

Dohrmann, U., Hertel, R., Pesci, P., Cocucci, S. M., Marré, E., Randazzo, G. and Ballio, A. (1977). *Plant Sci. Lett.* **9,** 291–299.

Dunkle, L. D. and Wolpert, T. J. (1981). *Physiol. Plant Pathol.* **18,** 315–324.

Earle, E. D., Gracen, V. E., Yoder, O. C. and Gemmill, K. P. (1978). *Plant Physiol.* **61,** 420–424.

Ellingboe, A. H. (1976). *In* "Physiological Plant Pathology", (R. Heitifuss and P. H. Williams, eds) pp. 761–778. Springer-Verlag Berlin, Heidelberg, New York.

Ellingboe, A. H. (1979). *In* "Recognition and Specificity in Plant Host-Parasite Interactions", pp. 3–17 (J. M. Daly and I. Uritani, eds) Japan Scientific Societies Press, Tokyo University Park Press, Baltimore.

Evans, M. L., Ray, P. M. and Reenhold, L. (1971). *Plant Physiol.* **47,** 335–341.

Federico, R., Scalia, S., Ballio, A., Cocucci, C., Ballarin-Denti, A. and Marré, E. (1980). *Plant Sci. Lett.* **17,** 129–134.

Fisher, R. A. (1968). *Science* **160,** 784–785.

Flor, H. H. (1955). *Phytopathology* **45,** 680–685.

Forde, B. G., Oliver, R. J. C. and Leaver, C. J. (1978). *Proc. Nat. Acad. Sci.* **75,** 3841–3845.

Frick, H., Nicholson, R. L., Hodges, T. K. and Bauman, L. F. (1976). *Plant Physiol.* **57,** 171–172.

Gardner, J. M., Mansour, I. S. and Scheffer, R. P. (1972). *Physiol. Plant Pathol.* **2,** 197–206.

Gardner, J. M. and Scheffer, R. P. (1973). *Physiol. Plant Path.* **3,** 147–157.

Gardner, J. M., Scheffer, R. P. and Higinbotham, N. (1975). *Plant Physiol.* **54,** 246–249.

Gilchrist, D. G. and Grogan, R. G. (1976). *Phytopathology* **66,** 165–171.

Giovanelli, I., Owens, L. D., Mudd, S. H. (1971). *Biochem. Biophys. Acta* **227,** 671–684.

Gracen, V. E., Grogan, C. O. and Foster, M. J. (1972). *Canad. J. of Bot.* **50,** 3167–3170.

Gregory, P., Earle, E. D. and Gracen, V. E. (1977). *In* "Host Plant Resistance to Pests", (P. A. Hedin, ed.) pp. 90–114. ACS Symposium series, No. 62.

Gregory, P., Matthews, D. F., York, D. W., Earle, E. D. and Gracen, V. E. (1978). *Mycopathologica* **66,** 105–112.

Gorshi, J., Solft, D., Shymala, G., Smith, D. and Hotedes, A. (1968). *Recent Prog. In Hormone Res.* **24,** 45–80.

Hanchey, P. and Wheeler, H. (1969). *Canad. J. of Bot.* **47,** 675–678.

Hanchey, P. and Wheeler, H. (1979). *In* "Recognition and Specificity in Plant Host–Parasite Interactions", pp. 193–210. (J. M. Daly and I. Uritani, eds) Japan Scientific Societies Press, Tokyo University Park Press, Baltimore.

Hanchey, P., Wheeler, H. and Luke, H. H. (1968). *Amer. J. Bot.* **55,** 53–61.

Harris E. J. and Pressman, B. C. (1967). *Nature* **216,** 918–920.

Haschke, H. P. and Luttge, U. (1977). *Plant Sci. Lett.* **8,** 53–58.

Hashimoto, T., Sakarai, A. and Tamura, S. (1967). *Pl. Cell. Physiol.* **8,** 23–34.

Hertel, R., St., Thomson, K. and Russo, V. E. A. (1972). *Planta* **107,** 325–340.

Hiroe, I., Nishimura, S. and Sato, M. (1958). *Tran. Tottori Soc. Agr. Sci.* **1,** 291.

Hoffman, S. E. and Zscheile, F. P., Jr. (1973). *Phytopathology* **63,** 729–734.

Hooker, A. L., Smith, D. R., Lim, S. M. and Beckett, J. B. (1970). *Plant Dis. Rep.* **54,** 708–712.

Ireland, C. R. and Strobel, G. A. (1977). *Plant Physiol.* **60,** 26–29.

Karr, A. L., Karr, D. B. and Strobel, G. A. (1974). *Plant Physiol.* **53,** 250–257.

Karr, D. B., Karr, A. L. and Strobel, G. A. (1975). *Plant Physiol.* **55,** 727–730.
Keck, R. W. and Hodges, T. K. (1973). *Phytopathology* **63,** 226–230.
Kenfield, D. S. and Strobel, G. A. (1980). *Biochim. Biophys. Acta* **600,** 705–715.
Kenfield, D. S. and Strobel, G. A. (1981a). *Plant Physiol.* **67,** 1174–1180.
Kenfield, D. S. and Strobel, G. A. (1981b). *Biochem. Internat.* **2,** 249–256.
Knudson, L. (1915). *Ann. Mo. Bot. Gard.* 2659–2666.
Koeppe, D. E., Cox, J. K. and Malone, C. P. (1978). *Science* **202,** 1227–1229.
Kohomoto, K., Scheffer, R. P. and Whitesite, J. O. (1979). *Phytopathology* **69,** 667–671.
Kono, Y. and Daly, J. M. (1979). *Bioorg. Chem.* **8,** 391–398.
Kono, Y., Knoche, H. W. and Daly, J. M. (1981). *In* "Toxins in Plant Disease", (R. Durbin, ed.) Academic Press, New York and London, (in press).
Kono, Y., Takeuchi, S., Kawarada, A., Daly, J. M. and Knoche, H. W. (1980). *Tetrahedron Lett.* **21,** 1537–1540.
Kono, Y., Takeuchi, S., Kawarada, A., Daly, J. M. and Knoche, H. W. (1981a). *Agr. Biol. Chem.* **44,** 2613–2622.
Kono, Y., Takeuchi, S., Kawarada, A., Daly, J. M. and Knoche, H. W. (1981b). *Bioorgan. Chem.* **10,** 206–218.
Kuo, Mong-shang, and Scheffer, R. P. (1970). *Phytopathology* **60,** 1391–1394.
Kuo, M. S., Yoder, O. C. and Scheffer, R. P. (1970). *Phytopathology* **60,** 365–368.
Lado, P., DeMichelis, M. I., Cerana, R., and Marré, E. (1976). *Plant Sci. Lett.* **6,** 5–20.
Lado, P., Rasi Caldogno, R. and Colombo, R. (1975). *Physiol. Plant* **34,** 359–364.
Lazarovits, G., Bhullar, B. S., Sugizama, H. J. and Higgins, V. J. (1979), *Phytopathology* **69,** 1062–1068.
Lazarovits, G. and Higgins, V. J. (1979). *Phytopathology* **69,** 1056–1061).
Lee, A. (1929). *Plant Physiol.* **4,** 193–212.
Lee, S., Auyagi, H., Shimohigashi, Y., Izumiya, N., Ueno, T. and Fukami, H. (1976). *Tetrahedron Lett.* **11,** 843–846.
Leonard, R. T. and Hodges, T. K. (1973). *Plant Physiol.* **52,** 6–12.
Lim, S. M. and Hooker, A. L. (1972). *Phytopathology* **62,** 968–971.
Litzenberger, S. C. (1949). *Phytopathology* **39,** 300–318.
Livingston, R. S. and Scheffer, R. P. (1981). *Phytopathology* **71,** 237 (abstract).
Luke, H. H. Warmek, H. E. and Hanchey, P. (1966). *Phytopathology* **56,** 1178–1183.
Maekawwa, N., Nishimura, S., Kohomoto and Watanabe, Y. (1979). *Annual Meeting of Japanese Phytopath. Soc.*, Program Issues II-23.
Maretzji, A. and Thom, M. (1978). *Plant Physiol.* **61,** 544–548.
Marré, E. (1979). Ann. Rev. *Plant Physiol.* **30,** 273–288.
Matthews, D. E., Gregory, P. and Gracen, V. E. (1978). *Phytopath. News* **12,** 162 (Abst. Ann. Meetings).
Matthews, D. E., Gregory, P. and Gracen, V. E. (1979). *Plant Physiol.* **63,** 1149–1153.
McCammon, S. L. (1978). *Phytopathology News* **12,** 162 (abstract).
McCammon, S. and Wheeler, H. (1978). *Physiol. Plant Pathol.* **12,** 191–198.

McFarland, B. L. and Gilchrist, D. G. (1981). *Phytopathology* **71,** 240 (abstract).
Meehan, F. and Murphy, H. C. (1947). *Science* **106,** 270–271.
Mertz, S. M., Jr. and Arntzen, C. J. (1972). *Plant Physiol.* **51,** 5–16.
Mertz, S. M., Jr. and Arntzen, C. J. (1977). *Plant Physiol.* **60,** 363–369.
Mertz, S. M., Jr. and Arntzen, C. J. (1978). *Plant Physiol.* **61,** 544–548.
Meyer, W. L., Kuyper, L. F., Lewis, R. B., Templeton, G. E. and Woodhead, S. H. (1974). *Biochem. and Biophys. Res. Comm.* **56,** 234–240.
Miller, R. and Koeppe, D. (1971). *Science* **173,** 67–69.
Moss, J. and Vaughn, M. (1979). *Ann. Rev. Biochem.* **48,** 581–600.
Nevers, P. and Saedler, H. (1977). *Nature* **268,** 109–115.
Nishimura, S., Kohmoto, K. and Otani, H. (1974). *RW. Plant Protect Res.* **7,** 21–32.
Nishimura, S., Kohmoto, K. and Otani, H. (1979). *In* "Recognition and specificity in plant host–parasite interactions", (J. M. Daly and I. Uritani, eds) pp. 133–146. Japan Scientific Societies Press, Tokyo.
Nishimura, S., Kohmoto, K., Otani, H., Fukami, H. and Ueno, T. (1976). *In* "Biochemistry and cytology of plant parasite interaction", (K. Tomiyama, J. M. Daly, I. Uritani, H. Oku, and S. Ouchi, eds) pp. 94–101. Kodansha Ltd., Tokyo and Elsevier, New York.
Nishimura, S. and Scheffer, R. P. (1965). *Phytopathology* **55,** 629–634.
Novacky, A. and Hanchey, P. (1974). *Physiol. Plant Pathol.* **4,** 1616–1675.
Odvody, G. N., Dunkle, L. D. and Edmunds, L., K. (1977). *Phytopathology* **67,** 1485–1489.
Ohkawa, M. and Torikata, H. (1967a). *J. Japan Soc. Hort. Sci.* **36,** 141–146.
Ohkawa, M. and Torikata, H. (1967b). *J. Japan Soc. Hort. Sci.* **36,** 263–267.
Okuno, T., Ishita, Y., Nakayama, S., Sawai, K., Fujeta, T. and Sawamura, K. (1974). *Ann. Phytopath. Soc. Japan* **40,** 375–376.
Okuno, T., Ishita, Y., Sawai, K. and Matsumoto, T. (1974). *Chem. Lett.* 635–638.
Okuno, T., Ishita, Y., Sugawara, A., Mori, Y., Sawai, K. and Matsumoto, T. (1975). *Tetrahedron Lett.* **5,** 335–336.
Otani, H., Nishimura, S. and Kohmoto, K. (1974). *Ann. Phytopathol. Soc. Japan* **40,** 59–66.
Park, P., Fukutomi, M., Akai, S. and Nishimura, S. (1976). *Physiol. Plant Pathol.* **9,** 167–174.
Pasika, A. M. and Cragg, L. H. (1963). *Can. J. Chem.* **41,** 293–299.
Payne, G. A. and Daly, J. M. (1978). *Phytopathology News* **62,** 162 (abstract).
Payne, G., Knoche, H. W., Kono, Y. and Daly, J. M. (1980a). *Physiol. Plant. Pathol.* **16,** 227–239.
Payne, G., Kono, Y. and Daly, J. M. (1980b). *Plant Physiol.* **65,** 785–791.
Pegg, K. G. (1966). *Queensl. J. Agr. Animal Sci.* **23,** 15–28.
Pham, H. N. and Gregory, S. P. (1980). *Plant Physiol.* **65,** 1173–1175.
Pressman, B. C. (1963). *In* "Energy-linked function of mitochondria", (B. Chance, ed.). Academic Press, New York and London.
Pressman, B. C. (1976). *Ann. Rev. Biochem.* **45,** 501–530.
Pinkerton, F. and Strobel, G. (1976). *Proc. Natl. Acad. Sci.* USA **73,** 4007–4011.

Pio Beltran, J., Strobel, G. A., Beier, R. and Mundy, B. P. (1980). *Plant Physiol.* **65,** 554–556.

Pringle, R. B. (1970). *Plant Physiol.* **46,** 45–49.

Pringle, R. B. (1971). *Plant Physiol.* **48,** 756–759.

Pringle, R. B. and Braun, A. C. (1957). *Phytopathology* **47,** 369–371.

Pringle, R. B. and Scheffer, R. P. (1963). *Phytopathology* **53,** 785–787.

Pringle, R. B. and Scheffer, R. P. (1964). *Annu. Rev. Phytopathol.* **2,** 133–156.

Pringle, R. B. and Scheffer, R. P. (1966). *Phytopathology* **56,** 1149–1151.

Pringle, R. B. and Scheffer, R. P. (1967). *Phytopathology* **57,** 530–532.

Rancillac, M. R., Sawney, R., Staskawica, B. and Galston, A. W. (1976). *Plant and Cell Physiol.* **17,** 987–995.

Raschke, K. (1975). *Ann. Rev. Plant Physiol.* **26,** 309–340.

Raven, J. A. and Smith, F. A. (1974). *Canad. J. Bot.* **52,** 1035–1048.

Raven, J. A. and Smith, F. A. (1976). *New Phytol.* **76,** 415–431.

Rawn, C. D. (1977). *Phytopathology* **67,** 338–343.

Rawn, C. D. and Van Etten, J. L. (1979). *Phytopathology* **68,** 1384–1388.

Rich, D. H. and Mathiapuraman, P. (1974). *Tetrahedron Lett.* **46,** 4037–4040.

Romanko, R. R. (1959). *Phytopathology* **49,** 32–36.

Saftner, R. A. and Evans, M. L. (1974). *Plant Pathysiol.* **53,** 382–387.

Saftner, R. A., Evans, M. L. and Hollander, P. B. (1976). *Physiol. Plant Pathol.* **8,** 21–34.

Sakai, R., Nishiyama, K., Ichihara, A., Shiraishi, K. and Sakamura, S. (1979). *In* "Recognition and Specificity in Plant Host–parasite Interactions" (J. M. Daly and I. Uritani, eds) pp. 165–179. Japan Scientific Societies Press, Tokyo.

Samaddar, K. R. and Scheffer, R. P. (1968). *Plant Physiol.* **43,** 21–28.

Samaddar, K. R. and Scheffer, R. P. (1970). *Plant Physiol.* **45,** 586–590.

Samaddar, K. R. and Scheffer, R. P. (1971). *Physiol. Plant Pathol.* **1,** 319–328.

Sawamura, K. (1966). *Bull. Hort. Res. Sta.,* Japan (Morioka) Series C, No. 4.

Scheffer, R. P. and Livingston, R. S. (1979). *Phytopathology* **70,** 400–404.

Scheffer, R. P. and Livingston, R. S. (1980). *Phytopathology* (in press).

Scheffer, R. P. and Pringle, R. B. (1961). *Nature* **191,** 912–913.

Scheffer, R. P. and Pringle, R. B. (1964). *Phytopathology* **54,** 832–935.

Scheffer, R. P. and Pringle, R. B. (1967). *In* "The Dynamic Role of Molecular Constituents in Plant–Parasite Interaction", (C. J. Mirocha and I. Uritani, eds) pp. 217–236. Bruce Publishing Co., St. Paul, MN.

Scheffer, R. P. (1976). *In* "Physiological Plant Pathology", (R. Heitefuss and P. H. Williams, eds) pp. 247–269. Springer-Verlag, Berlin.

Scheffer, R. P. and Samaddar, K. R. (1970). *Recent Advances in Phytochemistry* **3,** 123–142.

Scheffer, R. P. and Ullstrup, A. J. (1965). *Phytopathology* **55,** 1037–1038.

Schertzr, K. F. and Tai, Y. P. (1969). *Crop Sci.* **9,** 621–624.

Selinger, Z., Eimenl, S. and Schramm, N. (1974). *Proc. Nat. Acad. Sci.* (USA) **71,** 128–131.

Shepard, J. F. Bidney, D. and Shahin, E. (1980). *Science* **208,** 17–24.

Silverstein, R. M., Bassler, G. C. and Morrill, T. C. (1974). *In* "Spectrometric

Identification of Organic Compounds", 3rd Edition. John Willey & Sons, New York.
Smedegard-Petersen, V. (1977). *Physiol. Plant Pathol.* **10,** 203–211.
Smedegard-Peterson, V. and Nelson, R. R. (1969). *Can. J. Botany* **47,** 951–857.
Steele, J. A. Uchytel, T. F., Bhatnagar, P. and Rich, D. H. (1976). *Proc. Nat. Acad. Sci. (USA)* **73,** 2245–2248.
Steele, J. A., Uchytel, T. F. and Durbin, R. D. (1978). *Biochem. Biophys. Acta* **54,** 136–141.
Steiner, G. W. and Byther, R. S. (1971). *Phytopathology* **61,** 691–695.
Steiner, G. W. and Strobel, G. A. (1971). *J. Biol. Chem.* **246,** 4350, 4357.
Strobel, G. A. (1973a). *J. Biol. Chem.* **248,** 1321–1328.
Strobel, G. A. (1973b). *Proc. Nat. Acad. Sci. (USA)* **70,** 1693–1696.
Strobel, G. A. (1974). *Proc. Nat. Acad. Sci. (USA)* **71,** 4234–4236.
Strobel, G. A. (1975). *Scientific American* **232,** 80–88.
Strobel, G. A. and Hapner, K. D. (1975). *Biochem. Biophys. Commun.* **4,** 151–156.
Strobel, G. A. and Kenfield, D. (1981). *In* "International Cell Biology 1980–81", (H. G. Schweiger, ed.) pp. 477–482. Springer, Berlin.
Strobel, G. A. and Steiner, G. W. (1972). *Physiol. Plant Pathol.* **2,** 129–132.
Sugiyama, N., Kashima, C., Yamamoto, M. and Mohri, R. (1965). *Bull. Chem. Soc.* Japan **38,** 2028.
Sugiyama, N., Kashima, C., Yamamoto, M. and Sugaya, T. (1966). *Bull. Chem. Soc.* Japan **39,** 1573–1577.
Sugiyama, N., Kashima, C. and Yamamoto, M. (1967). *Bull. Chem. Soc.* Japan **40,** 345–346.
Tanaka, S. (1933). *Mem. Coll. Agri. Kyoto Imp. Univ.* **28,** 1–31.
Templeton, G. E. (1972). In "Microbial toxins" vol. VIII, (S. Kadis, A. Ciegler and S. J. Ajl, eds) pp. 169–192. Academic Press, New York and London.
Tipton, C. L., Mondal, M. H. and Benson, M. J. (1975). *Physiol. Plant Pathol.* **7,** 271–286.
Tipton, C. L., Mondal, M. H. and Uhlig, J (1973). *Biochemical and Biophysical Research Communications* **51,** 725–728.
Tipton, C. L., Paulsen, P. V. and Betts, R. E. (1977). *Plant Physiol.* **59,** 907–910.
Turner, N. C. (1972). K^+ uptake of guard cells stimulated by fusicoccin. *Nature*, London **235,** 341–342.
Turner, N. C. (1972). *Am. J. Bot.* **59,** 133–136.
Turner, N. C. and Grantiti, A. (1969). *Nature* **233,** 1070–1071.
Ueno, T, Hayashi, Y., Nakashima, T., Fukami, H., Nishimita, S., Kohmoto, K. and Sekiguchi, A. (1975a). *Phytopathology* **65,** 82–83.
Ueno, T., Nakashima, T., Hayashi, Y. and Fukami, H. (1975b). *Agr. Biol. Chem.* **39,** 1115–1122.
Ueno, T., Nakashima, T., Hayashi, Y. and Fukami, H. (1975c). *Agr. Biol. Chem.* **39,** 2081–2082.
Van Etten, J. L. and Daly, J. M. (1981). *Phytopathology* **70,** 727–729.
Walton, J. D., Earle, E. D., Yoder, O. C. and Spanswick, R. M. (1979). *Plant Physiol.* **63,** 806–810.

Watrud, L. S., Baldwin, J. K., Miller, R. J. and Koeppe, D. E. (1975). *Plant Physiol.* **56,** 216–221.

Wheeler, H. (1969a). *In* "Symposium on potential in crop protection", pp. 9–13. Cornell University.

Wheeler, H. (1969b). *Phytopathology* **59,** 1093–1097.

Wheeler, H. (1975). *In* "Plant Pathogenesis", 1–105 pp. Springer Verlag, Heidelberg.

Wheeler, H. (1976). *In* "Specificity in Plant disease", (R. K. S. Wood and A. Graniti, eds) pp. 217–230. Plenum Publishing Corp, N.Y.

Wheeler, H. and Black, H. W. (1962). *Science* **137,** 983–984.

Wheeler, H. and Uoupnik, B., Jr. (1969). *Phytopathology* **59,** 1460–1463.

Wheeler, H. and Hanchey, P. (1966). *Science* **154,** 1469–1571.

Wheeler, H. and Luke, H. H. (1963). *Ann. Rev. Microbiol.* **17,** 223–242.

Whiteside, J. O. (1976). *Plant Dis. Rep.* **60,** 326–329.

Wolpert, T. J. and Dunkle, L. D. (1980). *Phytopathology* **70,** 872–876.

Yoder, O. C. (1972). *In* "Phytotoxins in plant diseases", (R. K. S. Wood, A. Ballio and A. Graniti, eds.) pp. 457–463. Academic Press, London and New York.

Yoder, O. C. (1973) *Phytopathology* **63,** 1361–1366.

Yoder, O. C. and Gracen, V. C. (1977). *Plant Physiol.* **59,** 792–794.

Yoder, O. C., Payne, G. A., Gregory, P. and Gracen, V. E. (1977). *Physiol. Plant Pathol.* **10,** 237–245.

Yoder, O. C. and Scheffer, R. P. (1969). *Phytopathology* **59,** 1954–1959.

Yoder, O. C. and Scheffer, R. P. (1973a). *Plant Physiol.* **52,** 518–523.

Yoder, O. C. and Scheffer, R. P. (1973b). *Plant Physiol.* **52,** 513–517.

4

CROWN GALL: A MULTIPOTENTIAL DISEASE

D. J. Merlo

Division of Biological Sciences

University of Missouri, Columbia, Missouri

I. Introduction
II. Current Concepts in the Study of Crown Gall
  A. Milestones in Crown Gall Science
  B. Biology of Crown Gall
  C. Molecular Biology of Crown Gall
III. The Future of Crown Gall Research
  A. Model System of Host-Pathogen Interaction at Many Levels
  B. Model System for the Study of Plant Cell Growth and Regulation
  C. Model System for Cancer Studies
  D. Model System of Microbial Evolution
  E. Model System for the Study of Prokaryotic Gene Expression in a Eukaryotic Cell
  F. Potential Vector for Plant Genetic Engineering
IV. Factors Contributing to Elucidation of the Crown Gall Mechanism
V. Acknowledgements
References

I. INTRODUCTION

The study of the phenomenon of crown gall has been an enterprise that has frustrated, amazed and amused scientists for over 75 years. Despite the large amount of effort which was devoted to elucidation of the mechanism of transformation of plant cells to the tumour state by the crown gall organism, only in the last ten years has it been possible to provide answers to a few of the most basic questions concerning the induction of these interesting neoplasias. The body of published literature concerning crown gall induction, tumour growth, physiology of both tumours and the tumour-inciting bacteria, and a myriad of other aspects of crown gall science is overwhelming, and to an uncomfortable extent either misleading, erroneous or at least contradictory. Therefore, to attempt a comprehensive literature review would be a waste of time for both the author and the reader. For the reader requiring more information, I suggest the recent short review by Drummond (1979), or the older, more comprehensive reviews of Kado (1976) and Lippincott and Lippincott (1975).

This discussion has three objectives, as follows:

(1) to provide an up-to-date (mid-1980) overview of selected published works which serve as guides to our current understanding of the biology and molecular biology of crown gall;

(2) to provide a rather biased, semi-historical review of the scientific and technological factors that facilitated the elucidation of the crown gall mechanism; and

(3) to speculate on the types of research that future investigators of the crown gall problem may pursue.

With these ends in mind, I hope the reader will accept this chapter not as a repository of vital crown gall data, but rather as a stimulus for thought and discussion concerning future rationales and approaches to the basic science of plant disease.

II. CURRENT CONCEPTS IN THE STUDY OF CROWN GALL

A. Milestones in Crown Gall Science

1. Proof of Bacterial Etiology

While crown gall was first mentioned by Aristotle, and Cavara (1897) reported the isolation of tumour-inducing bacteria from diseased grapevines, most reviews of the history of crown gall recognize the works of Smith and Townsend (1907) as the definitive experiments proving the bacterial etiology of crown gall. The causal organism, *Agrobacterium tumefaciens*, is a Gram-negative, soil-borne pathogen ubiquitous to all cultivated areas of the world. Virtually every dicotyledonous plant is a susceptible host, and the list of affected plants comprises 142 genera classified in 61 families. This extremely broad host range, which includes no monocotyledons, is unique to the crown gall organism. Although some host specificity of bacterial strains has been found, there is not a single instance, to my knowledge, for a dicotyledon that is completely resistant to infection by some strain of the pathogen. This widespread susceptibility perhaps points to a set of basic biochemical or physiological processes common to dicotyledons but not present in monocotyledons. These render the plants' defence mechanisms, which are effective in resisting other microorganisms, inconsequential in fending off the crown gall organism. Alternatively, perhaps the mechanism of disease induction in crown gall is so unique that consideration of other disease resistance mechanisms is simply not appropriate.

2. Proof of Persistence of the Transformed State in the Absence of Bacteria
Besides its wide host range, the crown gall disease stands unique among plant diseases in that the principal symptom, the uncontrolled growth of the plant cells comprising the crown gall tumour, does not require the continued presence of virulent bacteria subsequent to the initial stages of tumour induction. This feature, first reported by Jensen (1910, 1918) and subsequently examined in greater depth by White and Braun (1941), defines crown gall neoplasias as true oncogenic tumours which are therefore distinguishable from the self-limiting overgrowths formed as insect galls or auxin-induced outgrowths that require the continued presence of the agent for symptom manifestation. The persistent aspect of the disease, allowing the unlimited axenic culture of tumour tissues, renders the crown gall system a perfect model for the study of many biological phenomena (see Section IV).

3. Proof of the Role of a Plasmid in Controlling Oncogenicity
During the several years since the proof of *A. tumefaciens* as the crown gall organism, virtually every component of the bacterial cells has been examined for a clue to the mechanism of tumour induction. Kado (1976) has well reviewed those efforts, and lists such diverse entities as bacterial L-forms, DNA, RNA, bacteriophages, ammonium ions, lipopolysaccharides, and endotoxin as subjects of examination by others in the search for the nature of the tumour-inducing principle (TIP) hypothesized some thirty years ago by Braun (1947).

The idea that *A. tumefaciens* elaborates a TIP grew directly from the previously mentioned observations that, once induced to express the tumour phenotype by a brief association with bacterial cells, the tumour cells remained as such in the absence of the bacteria. The rapid progress in molecular biology from 1950 onwards clearly gave much impetus to the notion that the TIP, which obviously exerted its influence in the plant cell long after the passage of viable bacteria, must be some sort of self-relicating entity such as DNA or RNA. It was not until the 1970s, however, that scientists were able to isolate and definitively characterize the nucleic acid which, first of all, confers to the bacterium the ability to induce tumours, and secondly, plays an indispensable role in tumour induction proper. This point will be further explained in a later section.

The works which contributed most directly to the discovery that a large plasmid (a covalently closed circle of episomal DNA) is responsible for conferring tumorigenicity to *A. tumefaciens* occurred in three separate laboratories. Kerr (1969) in Australia reported first that the tumour-inducing ability could be transferred *in planta* from a virulent to a non-virulent bacterium. Hamilton and Fall (1971) subsequently found that certain *A. tumefaciens* strains could be permanently cured of tumorigenic activity by non-mutagenic methods (viz. growth at 36°). These observations suggested the involvement of a dispensable, transferable genetic agent, and culminated in the report by Zaenen *et al.* (1974) that all virulent *A. tumefaciens* isolates examined contained a large (150–230 thousand base pairs; kbp) plasmid, whereas none of the non-virulent strains tested contained large plasmids. Clearly a correlation between plasmid content and tumorigenicity existed, but the final proof of the involvement of the plasmid came from the works of Watson *et al.* (1975) and Van Larebeke *et al.* (1975). Besides showing that heat-cured, non-virulent *A. tumefaciens* cells lacked detectable plasmid that had been present before the heat treatment, these workers also showed that all known tumorigenic characters were regained upon reacquisition of the plasmid by the cured bacterial cell. This final point is critical. The correlation between lack of plasmid and lack of a particular phenotype is interesting and intriguing, but proves nothing. The definitive proof of the involvement of a plasmid (or for that matter, any genetic element) in controlling some function must include evidence that the plasmid can confer the function to a cell that normally lacks that property. It is precisely this latter type of proof which is lacking in current literature reports of suspected involvement of plasmids in other types of plant disease (e.g. Gonzalez and Vidaver, 1979; Gantotti *et al.*, 1978). Since efficient gene transfer systems are lacking in most plant pathogenic bacteria, the development of such mechanisms should be a primary goal in the study of any plant disease at the molecular genetic level.

4. *Proof of Transfer of Plasmid DNA to Plant Tumour Cells*

As will be elaborated in Section II.A.6, much circumstantial evidence was available by 1975 which suggested that genetic information was passed, during the process of tumour initiation, from the bacterial cell to the plant cell. Faced with the insurmountable evidence that a plasmid (by that time, called the Ti plasmid, for "tumour-inducing") was intimately involved in conferring to the bacterium the capacity to induce plant tumours, it was only logical that scientists would devise experiments to test the possibility that the genetic information transferred to the plant cell was, in fact, all or a portion of the Ti plasmid. That this was indeed the case was first shown definitively for a tobacco tumour tissue by scientists at the Unitiversity of Washington in Seattle (Chilton *et al.*, 1977). An earlier report (Matthysse and Stump, 1976) that claimed detection of large amounts of plasmid sequences in *Vinca rosea* tumour tissue DNA was not corroborated by experiments performed with the same tissues in another laboratory (D. Sciaky, unpublished observations).

The report by Chilton *et al.* (1977) that tumour tissues contain Ti plasmid sequences has by now been well verified in at least three independent laboratories utilizing a tissue line that is a sibling to that examined by Chilton *et al.* (Gurley *et al.*, 1979), sunflower tumour tissues (ibid.), or independently derived tobacco tumour tissues (Ledeboer, 1978; Schell *et al.*, 1979). There can remain no doubt that the crown gall system does, in fact, represent a system wherein DNA sequences that normally reside in a prokaryotic cell are transferred to, maintained in, and expressed (see below) in the eukaryotic plant cell. For this reason alone, the crown gall system is unique.

5. Proof of Expression of T-DNA

At least two models have been suggested to explain the transformation to the uncontrolled growth state that occurs in plant cells upon acquisition of Ti plasmid DNA. In the first model, which I call the direct mechanism, the plasmid DNA itself serves as the controlling element that, after integration near the appropriate plant genes, destroys the normal control mechanisms of those genes. The results of this loss of control could be, for example, the constitutive synthesis of phytohormones (auxin and cytokinin), whose metabolism is so carefully controlled in the normal plant cell. A prokaryotic model of the direct mechanism exists in the insertion element IS2 (Saedler *et al.*, 1974). This small (1.33 kbp) piece of DNA has the ability, when integrated into the bacterial genome in the correct orientation, to promote the constitutive expression of adjacent genes. There is, however, increasing evidence that the direct mechanism is not a viable model for the crown gall system.

The second model, which I call the indirect mechanism, suggests that it is not the plasmid DNA *per se* that induces the transformed state, but rather an expression product (RNA, protein or both) that is responsible for altering the metabolism of the plant cell. Evidence in support of this model, has, so far, been circumstantial. Drummond *et al.* (1977) first demonstrated that RNA sequences homologous to a portion of the Ti plasmid found in tumour cells (i.e. the T-DNA) were present in tobacco tumour cells. These results were corroborated by Lederboer (1978), Gurley *et al.* (1979) and Yang *et al.* (1980a).

Recently, McPherson *et al.* (1980) have isolated from tobacco tumour tissues the RNA species that are complementary to the T-DNA, and have shown that these RNAs are capable of directing cell-free synthesis of proteins in a wheat germ system. It is apparent that these specific RNAs, although derived from prokaryotic DNA sequences, are treated by the plant cell as normal eukaryotic messenger RNAs. Support for this conclusion is provided by the facts that the procedure for purifying these RNAs included a step that enriches for species containing polyadenosine (poly-A), and the ability of the RNAs to direct protein synthesis was inhibited by compounds known to interfere with translation of capped eukaryotic messenger RNAs. Thus, we can infer that the RNAs contain eukaryotic-like 5′ cap structures, and 3′ poly-A tails, two characteristics that distinguish eukaryotic messenger RNA from the prokaryotic counterpart.

McPherson *et al.* (1980) found that only a few small (MW = 15 000, 16 500, and 30 000) proteins were specified by these tumour-specific RNAs. While the physiological functions of these proteins are unknown at present, it is tempting to speculate that at least one of them is responsible for induction and maintenance of the transformed state, analogous to the T-antigen produced by the mammalian SV-40 transforming virus (Kelly and Nathans, 1977). In any case, it is well established now that crown gall tumour cells contain DNA sequences that derived during the course of infection from a plasmid borne by the inciting bacterium, and that these DNA sequences are transcribed into eukaryotic-like messenger RNAs which are capable of supporting protein synthesis in a plant-cell-derived system. It is presumably only a matter of time before evidence proving the translation of these messenger RNAs in the tumour cell itself will be forthcoming.

6. *Correlation of Opine Production/Utilization and Bacterial Plasmid Content*

Circumstantial evidence suggesting the transfer of genetic information from *A. tumefaciens* to the crown gall tumour cell was published in 1970 by Petit *et al.* Their evidence consisted of a correlation between the type of tumour-specific compound formed in various crown galls, and the identity of the bacterial strain that had incited the particular tumour. The tumour-specific compounds, now collectively called opines, that were first discovered were octopine and nopaline. One or the other of these compounds, both arginine derivatives, was found to be present both in tumours excised from the plant, and in tumour tissue cultures that had been maintained axenically for many years. The identity of the compound was dependent not on the species of host plant infected, but on the identity of *A. tumefaciens* strain inoculated into the plant.

Furthermore, it was found that if a particular bacterial strain induced a tumour that contained octopine, those bacteria could utilize octopine very well as a carbon and/or nitrogen source, but nopaline was utilized either not at all, or very poorly. A similar situation exists for nopaline-utilizing strains and the tumours they induce. One possibility suggested by these correlations was that bacterial genetic information, which normally encoded an enzyme concerned with opine catabolism in the bacterial cell, was transferred to, and expressed in, the plant tumour cell. In this foreign biochemical environment, the enzyme was hypothesized to function only in the reverse, synthetic, direction. It therefore catalysed opine production. As attractive as this hypothesis was, it has subsequently been found to be incorrect, since the two enzymes (synthetic and catabolic) are distinguishable both biochemically and genetically (Drummond, 1979).

Although incorrect in detail, the model has been proven to be true in concept. It is now known that the genes that control opine utilization in the bacteria, and the genes that specify opine production in the tumour cell, are both borne by the Ti plasmid (see Section II.C.3). Furthermore, it has been shown that part of the T-DNA of the Ti-plasmid (i.e. the portion maintained in crown gall cells) includes the genes that specify opine production by the tumour cell. While it has not yet been shown that the enzyme(s) responsible for opine synthesis *per se* (perhaps one of the proteins found by McPherson *et al.* (1980)) are encoded by these genes, it is certainly clear that mutation or deletion of these Ti plasmid DNA sequences destroys opine synthesis in the plant cell. Because opines are not found in detectable amounts in normal plant cells (although there is still some debate on this point), it seems likely that the Ti plasmid genes do code for the opine synthesis enzymes, rather than derepress normally silent plant genes.

It should be noted here that the Ti plasmids can be classified into groups based on their ability to specify metabolism of a particular opine. Octopine type plasmids enable bacteria to catabolize octopine and induce tumours that synthesize octopine. Nopaline type plasmids specify the analogous nopaline function, and cannot coexist in the same bacterial cell as an octopine type plasmid. A third type of opine will be discussed later.

7. *The Concept of Genetic Parasitism*

One of the features of the field of plant pathology that I find most stimulating is the ever-present potential to discover a biological system that is totally new. The study of crown gall has now provided us with an example of a hitherto undescribed biological interaction which has been

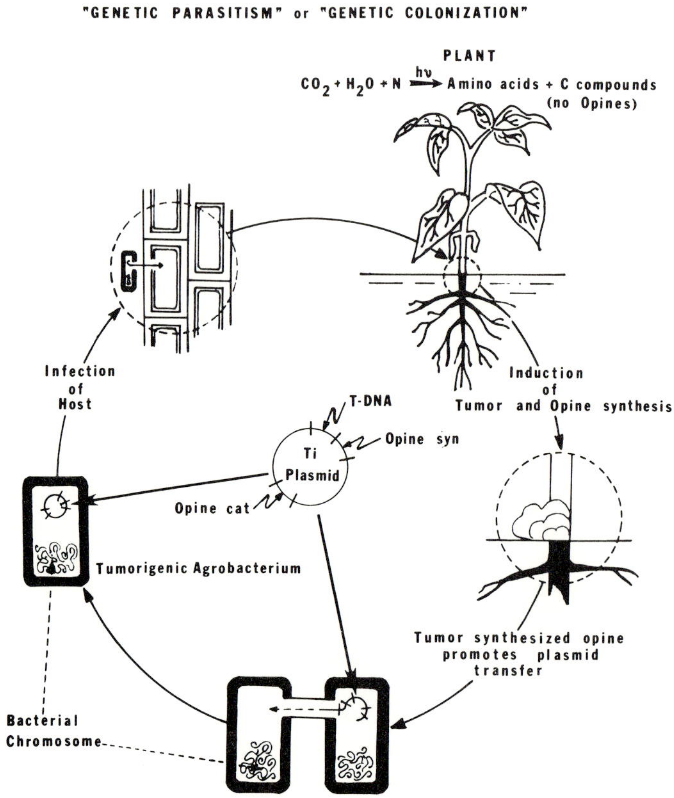

Fig. 1. Crown Gall represents the only known example of genetic colonization or parasitism. By altering the plant's genome through the transfer of bacterial DNA to the plant, the inciting organism induces the plant to produce a microenvironment favourable to the growth of the pathogen.

called genetic colonization by Schell (Rijksuniversiteit, Gent), and genetic parasitism by Nester (University of Washington, Seattle).

As we have seen, virulent *A. tumefaciens* cells harbour a large Ti plasmid that confers to the bacterium the ability to induce plant tumours. As part of this tumour induction process, a portion (less than 10%, the T-DNA) of the plasmid is transferred to, maintained in, and expressed in the plant cell. One of the results of this DNA transfer is the production by the plant tumour cell of specific compounds called opines. These opines, which have no apparent role in tumour cell growth (fully autonomous tumours can be found that contain none of the common opines), are able to be utilized as carbon and nitrogen sources only by the infecting bacteria that carry the Ti plasmid. Therefore, by inserting the appropriate genes into the plant cell, the bacterium has harnessed the plant's photosynthetic capability, causing the plant to produce a milieu favouring the pathogen's growth. Further, this milieu serves to encourage the spread of the pathogenic principle. The opines, it has been found (Kerr *et al.*, 1977; Genetello *et al.*, 1977), can also induce the conjugative transfer of the Ti plasmid to plasmidless strains of *A. tumefaciens*. In the presence of opines, the frequency of plasmid transfer was found to increase by four orders of magnitude.

This feature assures that any non-pathogenic agrobacteria which happen to be present in the wound site have a good probability of receiving the Ti plasmid, and therefore of becoming pathogenic. As Drummond (1979) has pointed out, this strategy could be considered an example of molecular altruism. Altruism is the name given to any behaviour of an individual which reduces that individual's fitness, while enhancing the reproductive potential of the population to which it belongs. In other words, *A. tumefaciens*, while increasing the competition for specific nutrients by transferring the genes that enable utilization of those nutrients to other bacteria that lack them, at the same time ensures that the Ti plasmid, and therefore the genes for production of those nutrients, is harboured by all competent members of the population. This enhances the survival probability of the Ti plasmid.

Taken together, these observations clearly serve as an example of a unique biological system. Figure 1 illustrates the concept, and is adapted from a drawing by Schell *et al.* (1979). Although the terms "genetic colonization" and "genetic parasitism" both refer to the same phenomenon, the latter term seems more descriptive of the actual process. It is not apparent that other examples of genetic parasitism will be found among plant pathogens, as some authors have suggested. Since crown gall serves as the only example of a plant disease that persists in the absence of the inciting agent (to my knowledge), perhaps it will serve also as the only example of this remarkable molecular interaction.

B. Biology of Crown Gall

1. The Infection Process

Much of our current knowledge of the early stages of development of crown galls comes from the extensive works of Armin Braun and co-workers. In a recent review (Braun, 1978), the tumour induction process was regarded as being comprised of two stages, the inception phase and the developmental phase. In the first phase, plant cells are converted to incipient tumour cells that do not yet undergo neoplastic growth, while the developmental phase is characterized by the persistent, uncontrolled growth of the tumour cells, even in the absence of viable bacteria.

The inception phase can itself be considered to comprise two developmental stages, conditioning and induction. Conditioned cells are those plant cells that have been rendered susceptible to transformation by the wounding process that is requisite for a successful crown gall infection. Biochemically, it is not known what features of the cell's metabolism are changed during conditioning, but what is known is that conditioning does not require the presence of *A. tumefaciens* cells, and that it is not an all-or-none process. Cells surrounding the wound site become increasingly more susceptible to transformation until 48–72 h after wounding. Before or after this period, tumours incited by inoculating the wound with virulent bacteria grow to a smaller size, and at a slower growth rate, than those incited at the optimum time. The fact that this period is just before the first cycle of cell divisions following wounding, and coincides with a period of rapid DNA synthesis, suggests the involvement of that process in genetically fixing the transformed state, but does not easily explain the reported differences in tumour size and growth rate. These differences have, incidentally, not been examined to any great extent since Braun's early work. Nor has the phenomenon of progression, the increasing susceptibility to transformation during conditioning, been adequately examined. It would be of great interest to learn what each of these terms means biochemically.

Physiological studies have revealed the presence of a thermosensitive step in the inception phase. In many plants, crown galls develop only if the temperature does not exceed 30°C during this stage. Once induced at lower temperatures, however, the tumours will grow well when placed at 30°C or above. Although temperatures in this range will cure some *A. tumefaciens* cells of their Ti plasmid, it is not likely that this can serve as the explanation for the lack of tumour induction at 30°C, since strains that are not heat-curable are nonetheless inhibited in tumour-inducing ability at this temperature. Tempé *et al.* (1977) found that the ability of the Ti plasmid to be transferred conjugatively between *A. tumefaciens* cells was also inhibited at 30°C or above, and suggested that perhaps a similar conjugation process was therefore involved between the bacterial and plant cells during tumour induction. If it is true, the genes controlling the process are either borne on the bacterial chromosome, or are as yet unmapped on the Ti plasmid, since no mutations on the Ti plasmid have been found which simultaneously affect tumour-inducing ability and bacterial conjugation of the Ti plasmid (Schell *et al.*, 1979).

Rogler (1980) has recently reported that, after tumour induction at

26°C, there follows a period of about seven days during which tumour growth rate is sensitive to elevated temperatures. After that time, the growth is not affected by high temperatures. Because tumours incited by *A. tumefaciens* cells that contain heat resistant plasmids become resistant to the effects of high temperature at earlier times, it is suggested that the heat sensitivity of the tumours is a plasmid-borne trait of undefined nature.

To summarize, much remains to be learned concerning the physiological and environmental factors that are involved in conditioning plant cells to transformation susceptibility. This topic must receive increasing research interest at many levels of experimentation if we are to attain a full understanding of the crown gall phenomenon.

Unfortunately, a similarly small amount of information is available concerning the biochemical events which transpire during the second stage of inception, the induction process. On a gross scale, we can say that inoculation of virulent agrobacteria into a wound site of a plant, under proper environmental conditions, results in the formation of a tumour ten days later. The exact nature of the interplay between plant and bacterial cell is almost completely unknown. The works of the Lippincotts (1969) have shown a probable requirement for binding of the bacterial cells to the plant cells, and others (Whatley *et al.*, 1978; Matthysse *et al.*, 1978) have provided evidence that the adherence specificity is a Ti-plasmid-controlled trait. Obviously, another event that must happen is the transfer of the T-DNA to the plant cell, but virtually nothing is known concerning this process. A basic question to be answered is whether the entire Ti plasmid is transferred to the plant cell, and then pared down to T-DNA size, or whether only the T-DNA itself is transferred. Marton *et al.* (1979) and Smith and Hindley (1978) have reported the transformation in culture of tobacco cells to the tumour state by bacteria, and perhaps this experimental system may be exploited to provide an answer to the above question. Another approach is to attempt *in vitro* transformation of plant cells (protoplasts seem a logical choice) by naked Ti plasmid DNA or plasmid fragments, and although rumours abound that such a feat has been accomplished, there is no published proof available to substantiate the claims. The use of liposomes to encapsulate the DNA, followed by fusion to plant cell protoplasts, may provide a means of effecting the transformation, since this technique is successful in introducing nucleic acids and other macromolecules into animal cells (Wilson *et al.*, 1979). At any rate, as Armin Braun has repeatedly stated (1978), the dual processes of conditioning and induction result in a transformed tumour cell that has acquired

> a capacity to synthesize persistently the essential growth-regulating substances and metabolites that its normal counterpart requires for cell growth and division but cannot make, at least not in amounts required to promote active cell growth and division.

This view maintains, therefore, that the tumour cell does not really differ biochemically from its normal counterpart except in quantity of metabolites, rather than identity. An extension of this thought predicts that normal cells, perhaps with the exception of the opine genes, contain all the genes that are derepressed in tumour cells, but are able to control their expression. Transformation, then, can be viewed as the loss of normal gene control, and this view may or may not prove to be correct. This point will be covered further in Section I.B.2.

The Developmental Phase. The temporal division between the induction and developmental phases of crown gall development corresponds to the phase of tumour growth after which the presence of the inciting bacterium is no longer necessary for the plant tissue to express the traits regarded as being characteristic of tumour cells. The numbers and identities of these tumour traits have been an area of controversy, but it is now generally agreed that the final criterion for determining whether or not a particular tissue is in fact a crown gall tumour is proof of the presence, and expression, in the candidate tissue of T-DNA sequences derived from an *A. tumefaciens* Ti plasmid. In the past, three characteristics that have been employed to distinguish tumour tissues from those derived from non-infected plants have been (i) the ability to grow in tissue culture in the absence of exogenous sources of two plant growth factors, auxin and cytokinin; (ii) the ability to express neoplastic growth when grafted to a healthy plant; and (iii) the presence of opines. These criteria, however, are not perfect, since some tissues derived from *A. tumefaciens*-induced tumours fail to display one or more of the criteria. For example, Braun (1958) reported that the transformation of a normal cell to a rapidly growing, fully autonomous tumour cell does not occur as an all-or-none phenomenon, but rather takes place over a three- to four-day period. Tumour tissue lines established early in this period have greater requirements for exogenous nutrients than those obtained at a later time. In addition, since tissues derived from non-infected plants can, with proper manipulation of their hormonal environment, be forced to grow without exogenous plant hormones, it is clear that nutritional requirements of a particular tissue line cannot serve alone as criteria to call that tissue a tumour. The other two criteria listed above likewise fail as universal characteristics of tumours.

It is true that, when grafted to a plant of high regenerative capacity such as tobacco, tobacco tumour tissues, but not normal tissues, will continue to grow. This growth differential is not seen, however, when the two tissue types from a species of low regenerative capacity, such as sunflower, are grafted onto healthy sunflower stems (Merlo, 1975). Finally, there are strains of *A. tumefaciens*, both mutant and wild type, that induce tumours that do not contain opines in detectable amounts (see Section I.B.2.b). This criterion, therefore, cannot be universally applied. As a closing comment, it should be noted that the criterion of the presence and expression of T-DNA is one that can be used only to define crown gall tumours, and workers studying other plant neoplasias such as wound tumour virus-induced overgrowths and Kostoff genetic tumours must devise their own sets of tumour definitions.

One of the most striking features of the developmental phase of crown gall growth is the appearance of different morphological types of tumours. These are generally observed as being of two classes, unorganized and teratoma. The latter tumour type is characterized by the production of a chaotic assembly of organs such as shoots and roots. The type of tumour resulting from a particular infection depends both on the species of host plant and the particular bacterial strain employed. Generally speaking, infection with octopine-type bacteria (see Section I.B.2.b) results in formation of unorganized tumours, whether the host plant be of high or low regenerative capacity. Infection with nopaline-type bacteria, on the other hand, can result in teratoma formation on some host plants (usually of high regenerative capacity), or in induction of unorganized tumours in other host plants. The response of the host plant to infection by a single strain can vary even among different species of the same genus, since strain T37 induces teratomas on *Nicotiana tabacum* and *N. langsdorffii*, but incites unorganized tumours on *N. sylvestris* and *N. glauca* (Gresshoff *et al.*, 1979). Since the determination of whether a particular bacterial strain is of the octopine or nopaline type is controlled by the Ti plasmid, it is not surprising that the tumour morphology incited by particular strains is also plasmid controlled (Gresshoff *et al.*, 1979).

Among the two general classes of tumours, there are also phenovariants that can commonly be found. In the unorganized tumours, one can find representatives that exhibit slow, moderate or rapid growth rates. Single cells isolated from teratomas can give rise to either an unorganized type of morphology, or to a morphology called "snowy" by Braun (1978). From these results, it is clear that crown gall tumours collectively represent not a single type of transformed cell, but rather include a multitude of cells and tissues that express a continuum of the degree to which the processes that control normal plant cell development have been aborted. This diversity of tumour types is one of the features that renders the study of crown gall so challenging, and information provided in the study of the molecular and physiological levels of control in the tumour may help us to understand the control of the normal cell.

The ability to regenerate whole plants from tissues derived from a single cell provides plant scientists with a valuable research tool not available in animal systems. In 1976, Braun and Wood, and Turgeon *et al.* reported the isolation of single cell clones derived from a teratoma tumour incited on Havana tobacco. Following grafting of this tumour tissue to the stem of Turkish tobacco plants, new shoots which were morphologically indistinguishable from Havana tobacco were initiated from the grafted tissue. The diploid tissues of these plants appeared normal in all characteristics examined (such as photosynthetic machinery), but reverted to the teratoma growth habit, with subsequent nopaline production, when placed in culture. The interpretation provided is that, when forced into rapid growth by the hormonal regime present at the plant wound site, the cells of the grafted tumour tissue were able to completely suppress the expression of some of the tumour phenotypes, although nopaline synthesis was sometimes not repressed (Wood *et al.*, 1978). When released from the normal plant's hormonal control by establishment in tissue culture, however, the expression of the tumour genes prevailed, and the tissues resumed the tumour growth habit. Recently Yang *et al.* (1980a, 1980b) examined tissue cultures established from one of these graft-induced, tumour-suppressed plants supplied by Braun. They found that tissue cultures derived from all diploid tissues of the plant contained T-DNA, but tissues derived from plant parts that had undergone meiosis (e.g. haploid anthers or plants established from F_1 seed) had lost the T-DNA. Clearly then, the normal-appearing, teratoma-derived plant still contained the genes specifying the tumour traits, but had suppressed the expression of those traits.

Reversal of tumour traits has also been reported by Sacristan and Melchers (1977) from single-cell clones of tobacco teratomas. However, since the source teratomas were incited by an *A. tumefaciens* strain that does not induce opine production, the absence of this biochemical marker leaves open the question of the true tumour identity of the individual single cells. This criticism is particularly valid in light of the fact that only three of the 341 clones established from the teratoma were phytohormone independent.

To conclude, the developmental phase of crown gall, as expressed by tissues grown in culture, rather than the induction phase, will probably be examined in greatest detail in the near future. This is due primarily to the ease of establishment and maintenance of single-cell-derived clones of tumour tissues that can be subcultured and scaled up in culture indefinitely. The preparation of protoplasts of tumour cells in the developmental phase will probably allow the genetic mapping of the tumour genes through fusion experiments with other cell protoplasts, while the preparation of subcellular components from large-scale homogenates of tumour tissue cultures should allow determination of the precise gene products encoded by the T-DNA. Whether or not these studies will bear a relationship to the events occurring in the healthy plant, or whether the elucidation of the cancer mechanism in plants has any bearing on animal cancers, remains to be seen, but clearly the answers to those questions cannot be attained without additional research.

2. *Types of Agrobacteria*

Taxonomy and Biotypes. Bergeys Manual (eighth edition) lists four species of the genus *Agrobacterium.* The taxonomic studies by Kersters *et al.* (1973) and others have been reviewed by Lippincott and Lippincott (1975). Most of the agrobacteria can be assigned to either of two biotypes based on physiological and biochemical characteristics, and recently organisms describing a third biotype have been isolated (Panagopoulos *et al.*, 1978; Kerr and Panagopoulos, 1977). In the author's opinion, the taxonomy of the agrobacteria is confusing at present, and a concerted effort to revise the taxonomy is overdue. Much of the problem stems from the assignment of species designations on the basis of pathogenicity and host range. This convention allows a cell bearing a particular bacterial chromosome to be classified as, for example, *A. tumefaciens*, if it harbours the C58 Ti plasmid (pTiC58); *A. radiobacter*, if it harbours no Ti plasmid (is avirulent); *A. rhizogenes*, if it harbours plasmid pHrAy from the hairy root organism; or *A. rubi*, if it can induce crown galls on *Rubus* species.

Anderson and Moore (1979) have dramatically pointed out the failings in the current classification system by showing that, for example, the hairy root organism *A. rhizogenes* induces tumours on some plants that are indistinguishable from the tumours induced on these plants by *A. tumefaciens*, whereas only *A. rhizogenes* induced hairy root formation on carrot disks *in vitro*. The classification of the organism as *A. rhizogenes* rather than *A. tumefaciens* would depend, in this case, on the judicious selection of appropriate host plants. In addition, *A. rubi*, classically considered to infect only *Rubus* species, was found to infect the same host plants as other strains of *A. tumefaciens*. Clearly, then, a taxonomy based on pathogenicity is not appropriate in this case.

The works of Anderson and Moore (1979) have also pointed out the quite narrow host range of some individual isolates of *Agrobacterium*, with most strains inducing tumours on six to eight host plant species (of 11 species tested). No strain was virulent on all 11 hosts. Work by others (Loper and Kado, 1979; Thomashow *et al.*, 1980b) has revealed that host specificity is plasmid-controlled in some instances, and further points out the absurdity of species designation by what essentially amounts to the cell's plasmid content. One resolution of this dilemma would be to adopt the proposal of Keane *et al.* (1970), and to consider all the strains currently classified as *A. tumefaciens* and *A. radiobacter* (and probably the other species as well) a single species (*radiobacter*), and then assign varietal designations as deemed fit (e.g. *A. radiobacter* var. *tumefaciens*).

Classification Based on Opine Metabolism. In much of the recent crown gall literature, it has been convenient to classify *Agrobacterium* strains as belonging to one of three opine types, that is, octopine strains, nopaline strains, or null strains. Bacteria of the first two types are able to utilize either octopine (not nopaline) or nopaline (not octopine) as the sole source of exogenous carbon or nitrogen in culture, and to induce tumours which produce either octopine or nopaline. The null-type strains were able to induce tumours that produced neither compound. The strain determination in each case, as mentioned previously, depends on the identity of Ti plasmid harboured by the particular bacterial cell. Faced with the close correlation of octopine or nopaline utilization and production evidenced by the majority of *A. tumefaciens* strains, researchers formulated a rationale termed "the opine concept" (Guyon *et al.*, 1980). This concept states that, since the production of opines by crown gall tumours apparently occurs to fulfil the nutritional needs of the oncogenic bacterium, then all crown gall tumour lines should produce opines. Although octopine and nopaline and their related compounds are both derivatives of amino acids and either pyruvic or α-keto glutaric acid (Table I), the biological properties of opines, inferred from the octopine and nopaline examples, should not depend on their chemical structure. In general, then, opines were defined (Guyon *et al.*, 1980) as

> a substance produced by the plant in response to a stimulus from the pathogen; its presence creates favorable environmental conditions for the pathogen and contributes to its dissemination.

Schell *et al.* (1979) have limited their definition to the specific crown gall

Table I. Opines Found in Crown Gall tumours.

Opine	Chemical name	Constituents	Reference
I. *Octopine Family*			
Octopine (oct)	{N^2-(D-l-carboxyethyl)-L-arginine}	Arginine+Pyruvic Acid	Ménage and Morel (1964)
Octopinic Acid (oca)	{N^2-(D-l-carboxyethyl)-L-ornithine}	Ornithine+Pyruvic Acid	Ménage and Morel (1965)
Lysopine (lop)	{N^2-(D-l-carboxyethyl)-L-lysine}	Lysine+Pyruvic Acid	Beimann *et al.* (1960)
Histopine (hop)	{N^2-(D-l-carboxyethyl)-L-histine}	Histidine+Pyruvic Acid	Kemp (1977)
Homooctopine (hoct)	{N^2-(D-l-carboxyethyl)-L-homoarginine}	Homoarginine+Pyruvic Acid	Petit and Morel (1966)
II. *Nopaline Family*			
Nopaline (nop)	{N^2-(1,3-dicarboxypropyl)-L-arginine}	Arginine+α-keto-Glutaric Acid	Goldmann *et al.* (1969)
Nopalinic Acid, or Ornaline (noa)	{N^2-(1,3-dicarboxypropyl)-L-ornithine}	Ornithine+α-keto-Glutaric Acid	Firmin and Fenwick (1977) Kemp (1978)
III. *Agropine*	$C_{11}H_{17}NO_7$	Amino Acid+Sugar	Firmin and Fenwick (1978)

case, stating that opines are (i) specifically synthesized by crown gall plant cells, (ii) synthesized either directly or indirectly under control of T-DNA from a Ti plasmid, (iii) able to be used selectively by agrobacteria as energy, carbon and/or nitrogen sources, and (iv) not necessarily, but usually, capable of inducing conjugative transfer of Ti plasmids.

Whichever of these definitions one used, the finding that tumours incited by the null-type strains, or other strains capable of nopaline utilization, contained neither octopine, nopaline, nor any of the members of their common chemical families, seemed at odds with the opine concept. Part of this dilemma was resolved recently by the finding (Guyon *et al.*, 1980) that the null-type tumours did contain an opine, but one whose chemical structure is completely unlike that of octopine and nopaline (Fig. 2; Coxon *et al.*, 1980). This new opine, agropine, was originally isolated as a compound which comprised almost 7% of the dry weight of an octopine-containing tumour tissue, and was found by Guyon *et al.* (1980) to be the predominant opine in tumours incited by a null-type strain. Since the null-type Ti plasmid employed in their study had a high degree of DNA base sequence homology with octopine type plasmids, it is not surprising that agropine was able to be utilized as a sole source of carbon and nitrogen by an octopine strain. Agropine differs biologically from octopine, however, in that it was ineffectual in inducing conjugational transfer of octopine type plasmids.

The only apparent exception to the opine concept, therefore, resides in the presence of nopaline-utilizing bacteria whose tumours contain no presently known opine. It will be of interest to determine if yet another opine type is present in those tumours, or if they indeed are the exception that proves the rule.

The apparent near-ubiquity of opines in crown gall tumours perhaps calls into question their presumed dispensability in the maintenance of the plant tumour cell growth habit. One view of the tumour cell maintains that the transformed cell has not suffered a drastic biochemical change in its makeup, but rather has lost the ability to control the synthesis of those components and systems that serve so well to control the growth of the normal cell. The appearance of tumour cells that lacked all of the known opines certainly gave credence to that view. The findings of Firmin and Fenwick (1978) and Guyon *et al.* (1980) that there also exists a completely new type of opine, which is present in tumour cells at levels far exceeding those of the conventional opines, again threw open the question of the possible involvement of opines in the maintenance of the tumour state. The pursuit of other members of the agropine class of opines is one which obviously should have great

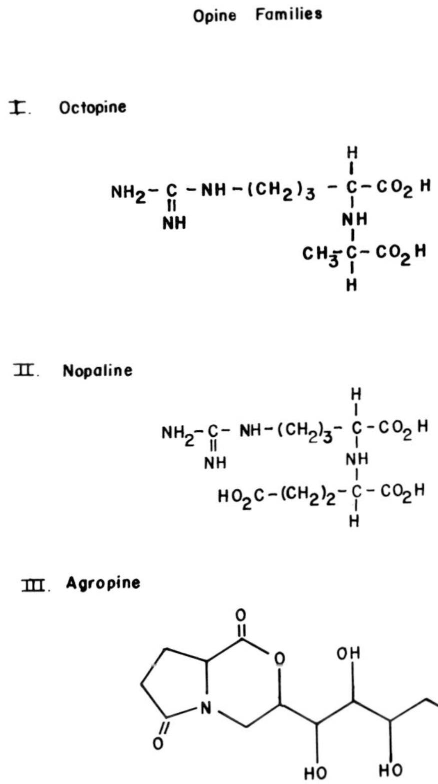

Fig. 2. Structures of the parent compounds of the three presently known opine families. The Octopine family consists of basic amino acids joined to pyruvic acid, whereas the Nopaline family consists of either arginine or ornithine joined to α-keto-glutaric acid. Agropine appears to be a bicyclic derivative of glutamic acid and a hexitol sugar. Table I presents a list of the members of each family.

appeal to scientists concerned with the biochemistry of the transformed plant cell.

The classification of the agrobacteria, as we have seen, can be considered on many levels, and as in any taxonomy, the field is in constant flux. Having established the tumorigenic potential of a strain, the further subclassification must rely on the aims of the investigator. It seems reasonable, however, in the light of recent work, to refer to the null type of plasmids and bacteria (first named by Sciaky *et al.*, 1978) now as the agropine type. This is more in accordance with the nomenclature of the octopine and nopaline types, and follows the suggestion of Guyon *et al.* (1980).

3. Biological Control of Crown Gall

Although this chapter is concerned mainly with the biochemistry and genetics of crown gall, it should not be forgotten that crown gall is a plant disease. Although recent figures are not available, the actual and potential losses of marketable plants in the nursery industry and reduction in yields in orchards infested with the disease total millions of dollars annually. In 1976, for example, crown gall was judged to have caused losses totalling $23 million in California alone (Kennedy and Alkorn, 1980). Primary losses occur in fruit and nut orchards and in other situations where extensive pruning and grafting, processes which provide a wound-site infection court, are necessary. In nurseries providing plant material for commercial growers, the most damaging infections occur at the final harvesting stage, when the trees are uprooted, root-pruned, and then offered for sale. Infections during these operations do not become evident until after the trees have been transported and planted in the orchard, and one infected tree, if transported in contact with non-infected material, may serve as an inoculum source to contaminate the entire lot. If the infected plants are ornamental varieties such as roses, crown galls destroy the marketability of the plants to both commercial and non-commercial buyers. Extensive crown galls on fruit trees, however, can decrease yields by restricting nutrient transport through infected stems and roots.

Control of the disease is possible at three levels. The first is to purchase plants from growers whose nurseries are free of the disease. Once infected, however, some measures may be taken to destroy the tumour tissue without damaging the plant extensively. Schroth and Hildebrand (1968) have described the eradication of crown galls on stone fruit and pear trees and other plants, through use of the commercial product Bacitin^R, a mixture of aromatic hydrocarbons in oil. Moore and Schroth (1976) also found that selectively toxic aromatic hydrocarbons were able to kill crown galls on tomato plants.

The most dramatic results, however, are obtained through prevention of the disease by counterinfection with strain 84 of *A. radiobacter*. This non-virulent strain was first used as a biological control agent by New and Kerr (1972), who found it to be the only one of seven avirulent strains tested capable of preventing crown galls. Moore (1977) found that by simply dipping young *Prunus* roots in water suspensions of strain 84 cells, natural crown gall infections were significantly reduced. This means of control has proved so effective that strain 84 is now commercially available as a crown gall preventative.

The mechanism of action of strain 84 in preventing crown galls is unknown. Kerr and Htay (1974) have suggested that the prevention is due primarily to the elaboration by strain 84 of a potent bacteriocin, rather than the possibility that strain 84 cells saturate all available binding sites on the plant cell walls, thereby preventing adherence of virulent cells. They found that seven of 14 strains tested were resistant to bacteriocin 84 and not subject to biocontrol by strain 84. The partial structure of agrocin 84, a 6-N-phosphoramidate of an adenine nucleotide analogue, has been published (Roberts *et al.*, 1977). The preventative effects of strain 84 are probably not totally due to the bacteriocin, however, as Moore (1977) found that the crown gall incidence caused by a mixture of virulent strains was reduced by treatment with strain 84, despite the fact that one of the virulent strains was resistant to bacteriocin 84 in *in vitro* tests. Smith and Hindley (1978) have reported that treatment of strain C58 cells with agrocin 84 significantly reduces the binding of bacterial cells to plant cells, and this feature may therefore play a role in biocontrol.

A final feature of the phenomenon of crown gall control by agrocin 84 concerns the fact that, in many strains, sensitivity to the bacteriocin is a trait borne by the Ti plasmid (Engler *et al.*, 1975). Indeed, one can select for bacterial cells cured of their Ti plasmid by selecting for resistance to the bacteriocin. This feature may also play a role in biocontrol, since those cured cells become avirulent with the loss of the plasmid. Murphy and Roberts (1979) have proposed that the sensitivity to agrocin 84 is due to the presence of plasmid-coded proteins which are associated with the transport of the bacteriocin into sensitive cells. One wonders if perhaps one or more of these hypothetical proteins corresponds to the periplasmic proteins found to be plasmid specified by Sonoki and Kado (1978).

C. Molecular Biology of Crown Gall

1. *Role of the Ti Plasmid in Controlling Tumorigenicity*

In a previous section, I discussed the evidence suggesting that the T-DNA probably plays an indirect role in inducing the tumour state of the transformed cell. Several aspects of this model bear closer examination.

At present, there is no clear evidence suggesting the biochemical nature of the T-DNA products which actually abort the normal control mechanisms of the plant. This loss of control probably occurs at several locations, and the genes specifying auxin and cytokinin production by the plant are obvious choices. It is possible, however, that at least one of these phytohormones may be a T-DNA product, since cytokinin production is thought by some investigators to be a Ti-plasmid-controlled trait (Messens and Claeys, 1978; Einset, 1980). Clearly, culture filtrates of *A. tumefaciens* cells contain cytokinins (Upper *et al.*, 1970), but there is still some debate as to the presence of unusual levels or identities of cytokinins in crown gall tissues (Braun, 1978).

One must also be concerned with the role of all the unusual compounds found in crown gall tissues. Although octopine was first isolated from marine molluscs, compounds such as nopaline and agropine have no known natural sources other than crown gall tumours. The absence of known opines from some tumour tissues does not preclude the involvement of other, yet to be discovered, compounds in inducing the persistent synthesis of compounds such as the phytohormones. It might be, moreover, that phytohormone-independent tumours represent only one type of cell that is present in the newly induced tumour. All current methods for culturing tumour tissues employ selection of the cells on medium lacking the phytohormones. Other cells, exhibiting a constitutive synthetic capacity for other compounds, may also be present but go undiscovered because they have not been looked for. For example, myoinositol and thiamine are often added to the selective medium (e.g. Linsmaier and Skoog, 1965). It is possible that cells expressing the persistent synthesis of those compounds may be found in a crown gall tumour. If so, then we would be forced to re-evaluate the role of phytohormones in expressing the tumour phenotype. The availability of auxotrophic mutants of the host plants would certainly facilitate this type of study.

Another feature worth mentioning is the apparent co-evolution of the separate traits of (i) the ability of *A. tumefaciens* DNA to induce the transformed state in plant cells, and (ii) the development of the delivery system to transfer this DNA to the plant cell. With the discovery that the Ti plasmid conferred to the bacterial cell the capacity to induce tumours, there was really no *a priori* reason to believe that the same piece of DNA would necessarily have any involvement in the induction of the transformed cell. One could have imagined, for example, a two-part system. The actual transformation could have been accomplished, for example, by a virus nuclei acid transferred to the plant cell. Genes on the Ti plasmid would have rendered the bacterial cell capable of replicating or transferring the virus. We now know, of course, that the genes affecting the transformed state are present on the Ti plasmid as the T-DNA. However there are some chromosomal mutations of the *A. tumefaciens* genome that also affect oncogenicity. Some of these mutations are merely auxotrophic in nature, and the loss of oncogenicity may be a pleiotrophic effect due to reduced fitness in the plant wound site (Lippincott *et al.*, 1965). Other strains, such as 5 Gly Fe (Sciaky *et al.*,

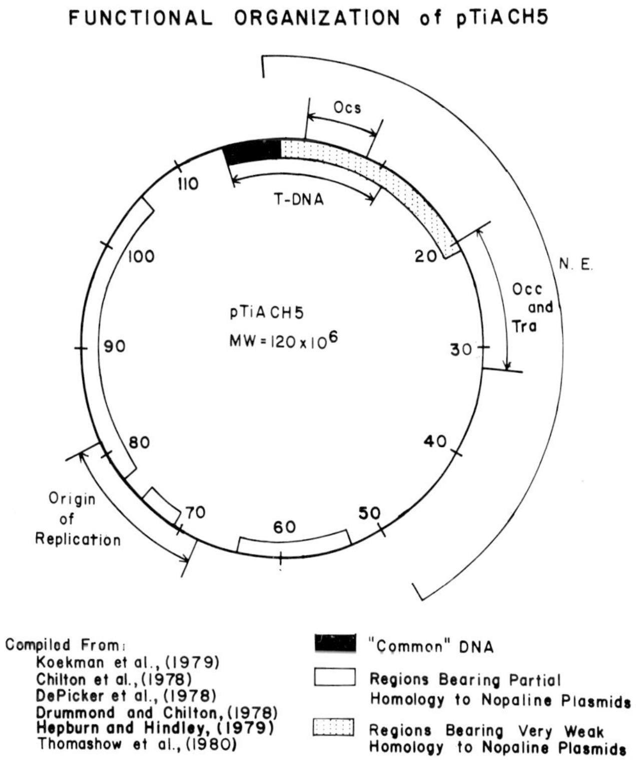

Fig. 3. Functional organization map of an octopine-type Ti plasmid, pTiACH5, which shows the physical relationship of known gene locations to areas of DNA sequence homology with a nopaline-type plasmid, and to the T-DNA. Most gene locations were determined by analyses of drug resistance transposon-generated mutations, or by deletion mutant analysis. N.E. denotes areas which are not essential for oncogenicity. Gene abbreviations are listed in Table II. Numbers refer to distance from the right end of the common DNA, in units of molecular weight $\times 10^6$.

1978) harbour a Ti plasmid that is capable of conferring oncogenicity to other strains, but the parent strain is prototrophic and non-tumorigenic. Davis and Rothman (1973) found that the proportion of non-tumourigenic and auxotrophic cells in nitrosoguanidine-treated cultures was similar, and that both cell types accumulated at about the same rate. If this suggests that there are as many cistrons required for virulence as can lead to auxotrophy, then there must be many genes, on the Ti-plasmid as well as the chromosome, which are essential for virulence. It will truly be a challenge to learn the identities of those gene products, and whether they comprise part of the Ti plasmid transfer system that has so far escaped elucidation. Perhaps the chromosomal gene transfer system developed by Hamada *et al.* (1979), or the transducing phage of Milani and Herberlein (1972) will be exploited in answering those questions.

A crucial question concerning the role of the T-DNA in tumorigenicity has been whether or not the T-DNA is covalently attached to plant DNA, whether this be nuclear or organellar in origin. Three papers have recently been published which address that issue. Thomashow *et al.* (1980a) studied the physical arrangement of the T-DNA in four independently-derived cloned tobacco tumour tissues that were incited by *A. tumefaciens* strains bearing octopine-type plasmids. Their data suggest that most (perhaps all) of the T-DNA is integrated into plant DNA, and that, although preferred regions of the Ti plasmid serve as points of attachment to the plant DNA, there is more than one integration site in the plant genome. Further, independent tumour lines were found to contain different amounts of T-DNA even though they were incited by bacteria bearing the same Ti plasmid. All tumour lines contained a core T-DNA, however, and this core contains the common DNA found on all Ti plasmids (see below). These results corroborate earlier work (Merlo *et al.*, 1980) that showed that independently derived tumours incited by bacteria bearing octopine type plasmids contained different extents of T-DNA, and the numbers of copies of different portions of the T-DNA varied amongst the tumours studied. This latter result suggested that the T-DNA is not a collinear copy of the DNA in the Ti plasmids since the right and left ends of the T-DNA were present in unequal copy numbers in the DNA from at least one tumour line. Further evidence of the non-collinearity of the T-DNA comes from the observation (Thomashow *et al.*, 1980a) that certain segments of the Ti plasmid which lie internally to the right and left ends of the T-DNA on the plasmid physical map were absent in some tumour tissue DNAs. Taken together, these results suggest that the portions of the plasmid maintained in tumour cells undergo rearrangements such as deletions and amplifications, and any model of the integration process must include these separate events.

Additional evidence revealing the complex state of the T-DNA in the strain T-37-induced teratoma line discussed in Section I.B.1.b has recently been published (Zambryski *et al.*, 1980). By use of recombinant DNA methods, these scientists have obtained cloned fragments of DNA isolated from the tobacco teratoma tissue. Analyses of these cloned DNAs suggest the following conclusions:

(i) The right end of the pTiT37 T-DNA (for our purposes,

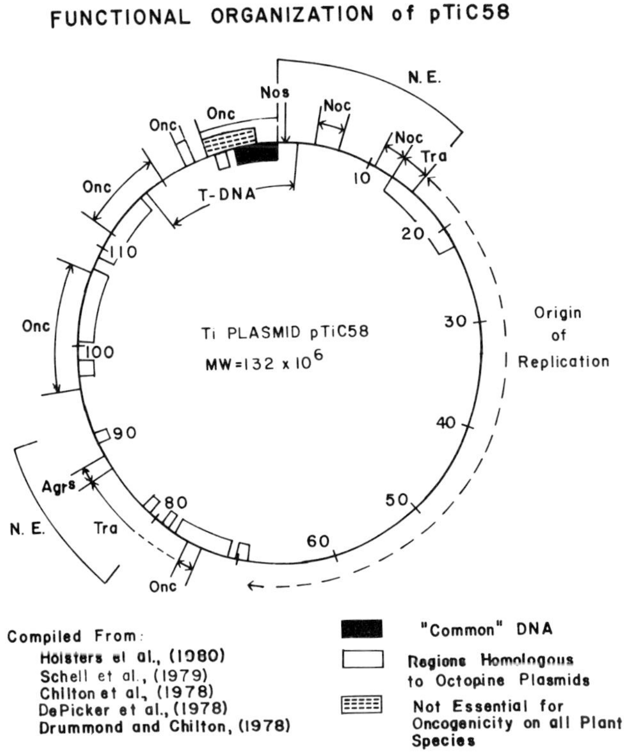

Fig. 4. Functional organization map of a nopaline-type Ti plasmid, pTiC58. For explanation of abbreviations and map units, see legend to Fig. 3.

considered to be identical to that shown in Fig. 4 for pTiC58) is attached to two kinds of DNA. One cloned fragment containing the right end also contains DNA sequences apparently derived from repeated sequences of tobacco DNA. This suggests the right end of the T-DNA is covalently attached to plant DNA that is present in many copies per tobacco genome. Another cloned fragment, however, reveals that the right end of the T-DNA is attached to the left end of the T-DNA, with a short piece of DNA of unknown origin being present between these two ends. This result suggests that multiple copies of the T-DNA are present in the cell as randomly repeated units. The second cloned fragment, therefore, is interpreted to have derived from the junction point where the right end of one T-DNA unit is attached to the left end of the next unit.

(ii) DNA sequences analysis of the DNA fragment present between the two T-DNA end fragments of the second clone reveals the presence of several direct repeats and one inverted repeat. These are features commonly found in transposable DNA segments in prokaryotes (Calos and Miller, 1980), and suggest the involvement of this intervening fragment in the integration process.

(iii) These workers were unable to isolate a DNA fragment in which the left end of the T-DNA was found to be attached to plant DNA. The question remains open, therefore, as to whether both ends of the T-DNA must be associated with repeated sequences of plant DNA.

(iv) The possibility that the T-DNA exists in the plant both as integrated and independently-replicating structures remains unresolved.

The physical location of the plant DNA to which the T-DNA is attached could be the nucleus, the mitochondrion, the chloroplast, or any combinations of these structures. The reversion studies discussed in Section I.B.1.b suggested that integration into organellar DNA might provide an easy explanation for reversion through meiotic segregation of the organelles. The evidence of Chilton *et al.* (1980), however, suggests that this is not a viable hypothesis. Their examination of the chloroplast and mitochondrial DNAs of the teratoma-derived, apparently normal plant described above detected no T-DNA sequences present in the organellar DNA, whereas substantial hybridization of the plasmid DNA to plant nuclear DNA was found. These workers also found evidence that the T-DNA was covalently attached to plant DNA, but as yet no evidence exists which bears on the question of integration into chromosomes *per se*, or rather the existence of a nuclear episome comprised of plant and plasmid DNA.

In any case, there is now enough evidence compiled to state with a good degree of certainty that the T-DNA is present in the nucleus in a manner which is more complex than a single simple insertion event, and that its transcription and the subsequent processing of the RNA are brought about by eukaryotic enzymes. The critical event catalysed by the presence of this foreign DNA that finally converts the normal cell to the tumour state is unknown, and will undoubtedly be the subject of much future research.

2. *Ti Plasmid-controlled Traits*

The proof of the plasmid-borne nature of a particular trait can only be complete if a series of experiments analogous to Koch's Postulates is performed. These should include at least the following: (i) demonstration that cells expressing the trait contain a plasmid; (ii) proof that cells not expressing the trait do not contain the plasmid (these cells preferably should be derivatives of the cells in (i) that have been cured of their plasmid by non-mutagenic methods); (iii) proof that cured cells acquire the trait upon reacquisition of the plasmid; and (iv) demonstration that mutations in the plasmid DNA affect the expression of the trait.

The availability of conjugation systems (Kerr *et al.*, 1977; Genetello *et al.*, 1977) and a transformation procedure (Holsters *et al.*, 1978), along with the use of plasmidless *Agrobacterium* strains, has enabled the determination of many genetic traits borne by the Ti plasmid. I have already discussed several of these traits; a more complete list is found in Table II. Not all of the traits that are listed in Table II and are suggested to be Ti-plasmid-controlled have been proven to be so by all the above criteria. Those whose plasmid-borne nature seems unequivocal have been assigned three-letter abbreviations; the final proof of the other traits has not yet been published. In those cases in which multiple publications are available which suggest the involvement of plasmid genes, I have tried to select the papers which offer the most complete proof to that effect.

In some cases, it should be noted that a particular plasmid-borne gene affects functions other than that for which the gene is named. For example, mutations in the octopine synthesis gene, *Ocs*, would eliminate synthesis of such compounds as histopine and octopinic acid, since these compounds presumably are products of the same enzyme responsible for octopine synthesis (Hack and Kemp, 1977). As will be seen in the following section, this list of plasmid-borne traits is far from sufficient to account for all the DNA sequences present in the Ti plasmid. Even if one allows a somewhat generous four or five million daltons of DNA per gene, we would still have accounted for only about one half of the plasmid genome. Obviously, further research will discover new traits borne by these interesting molecules.

3. Physical and Genetic Maps of Ti Plasmids

Early work by Currier and Nester (1976) revealed that on the basis of DNA hybridization studies, all octopine-type Ti plasmids were nearly 100 % homologous to one another. This high degree of evolutionary conservation did not extend, however, to the nopaline-type plasmids, which were found by Currier and Nester (1976) and by Drummond and Chilton (1978) to contain 5–35% homology with the octopine-type plasmids. Hepburn and Hindley (1979) found that the areas of homology between an octopine and a nopaline-type plasmid were not clustered, but instead were scattered on the octopine-type plasmid physical map. Sciaky *et al.* (1978) had earlier shown that, in agreement with the DNA homology studies, all octopine-type plasmids had virtually identical restriction endonuclease fingerprints. This canonical

Table II. Ti plasmid-determined characteristics.

Trait	Abbreviation[a]	Reference
(1) Oncogenicity	Onc	Van Larebeke *et al.* (1975) Watson *et al.* (1975)
(2) Bacterial octopine or nopaline catabolism (requires permease and oxidase)	Occ or Noc	Van Larebeke *et al* (1975) Bomhoff *et al.* (1976) Montoya *et al.* (1977)
(3) Tumoral octopine or nopaline synthesis	Ocs Nos	Bomhoff *et al.* (1976) Montoya *et al.* (1977)
(4) Arginine catabolism	Arc	Ellis *et al.* (1979)
(5) Ornithine catabolism	Orc	Holsters *et al.* (1980)
(6) Tumoral production of agropine		Firmin and Fenwick (1978)
(7) Tumour morphology (teratoma or unorganized)		Gresshoff *et al.* (1979)
(8) Sensitivity to agrocins 84 and 396	Agr^s	Van Larebeke *et al.* (1975) Engler *et al.* (1975)
(9) Autotransfer to bacterial cells	Tra	Kerr *et al.* (1977) Genetello *et al.* (1977)
(10) Exclusion of phage API	Ape	Van Larebeke *et al.* (1975)
(11) Adherence to plant cell walls		Whatley *et al.* (1978) Matthysse *et al.* (1978)
(12) Periplasmic proteins per-1 and per-2		Sonoki and Kado (1978)
(13) Plasmid incompatibility		Montoya *et al.* (1978) Genetello (1978)
(14) Plant host range		Loper and Kado (1979) Thomashow *et al.* (1980b)
(15) Cytokinin production		Messens and Claeys (1978) Einset (1980)
(16) Indoleacetic acid production		Liv and Kado (1979)
(17) Restriction enzyme Atu BVI		Roizes *et al.* (1979)

[a] When appropriate, the phenotype abbreviation, as agreed upon at the IVth International Conference on Plant Pathogenic Bacteria (Angers, France, August 27 to September 1, 1978) is given.

fingerprint for octopine-type plasmids differed markedly from those of the nopaline-type plasmids. Further, the fingerprints (Sciaky *et al.*, 1978) and DNA homology studies (Currier and Nester, 1976) revealed that the nopaline-type plasmids are themselves an evolutionarily diverse group, having as little as 67% homologous sequences, and quite dissimilar restriction endonuclease fingerprints. These results may suggest that the nopaline-type plasmids are evolutionarily much older than the octopine-type plasmids, and therefore have diverged to a significant extent.

Genotypically, the octopine and nopaline-type plasmids share several traits. Both are able to induce plant tumours, both transfer plasmid DNA to the plant cell, both are autotransmissible, both can replicate in *Agrobacterium*, and bacteria harbouring either type of plasmid must adhere to plant cells in order to induce tumour formation (Table II). One might expect, therefore, that a genetic map of these traits would coincide with a physical map depicting areas of DNA base sequence homology, and to a certain extent this prediction is true. In Figs 3 and 4, the functional maps of an octopine and a nopaline plasmid are presented, and the coincidence of the locations of traits such as oncogenicity with the positions of DNA base sequence homology can easily be seen. To align these maps with one another, the end of the most highly conserved portion of the plasmids (see below) which lies in the T-DNA, to the left of the genes governing opine synthesis, is taken as the starting point. Several features of the maps deserve additional comment.

(a) There is a region of DNA, the common DNA, that is highly conserved among all Ti plasmids that have been studied (Chilton *et al.*, 1978; DePicker *et al.*, 1978). The common DNA, comprising about 5 million daltons (8·1 kbp) is included in the T-DNA of both octopine and nopaline-type plasmids, and mutations in the common DNA destroy the plasmid's ability to induce tumours. The size of the T-DNA in Fig. 3 represents that found in the tobacco tumour lines E9 and 15955/1 as shown by Thomashow *et al.* (1980). These extents of T-DNA are for diagrammatic purposes, since, as mentioned above, the exact extent of T-DNA is found to be different among different tumour lines. In all cases reported, however, the T-DNA includes the common DNA, and this suggests a common role for this DNA in tumour induction and/or maintenance.

(b) There are other regions of homology between octopine and nopaline-type plasmids scattered around the physical maps of the plasmids. In most cases, these areas are less highly conserved than the common DNA, having thermal melting midpoints somewhat lower than that of common DNA. Particularly in the case of pTiC58 (Fig. 4), it can be seen that the *Tra* and *Onc* traits have been mapped in these areas of homologous DNA. This implies that these genes share a common hereditary ancestor, and that similar mechanisms are operative in expressing those traits.

(c) The genes controlling opine synthesis are located within the T-DNA, whereas the genes controlling opine catabolism occur outside the T-DNA region. Besides providing additional evidence for the genetic independence of these two traits, these observations provide a teleological argument to explain the frequent occurrence of opine production in tumour cells. One might argue, for example, that fewer tumours which produce opines might be isolated if the genes controlling opine production occurred at a more distant map position from the common DNA. This argument assumes a primary role of the common DNA in determining the extent of the plasmid which is transferred to, or maintained in, the tumour cell.

(d) Mutations in several map positions are capable of destroying oncogenicity. Although no functions have as yet been ascribed to the various *Onc* regions, one might expect that those found to map outside the T-DNA region could control the adherence to the plant cell wall, and transfer of the plasmid DNA to the plant cell. With regard to the latter trait, it is perhaps significant that no mutations have been found that simultaneously destroy oncogenicity and the transfer of the plasmid to other bacterial cells. This indicates that the latter trait is probably separate from the bacteria-to-plant cell transfer mechanism.

(e) The origins of replication of pTiAch5 and pTiC58 overlap in homologous regions. The plasmid incompatibility phenomenon (Table II) has been suggested to be a result of competition of the plasmids for a common binding site on the bacterial cell membrane. The presence of homologous sequences in the replication origins adds credence to that hypothesis. It seems possible, however, that a plasmid-encoded restriction endonuclease (Table II) could also be responsible for the incompatibility of octopine and nopaline type plasmids.

To summarize, we have learned a great deal about the evolutionary conservation and genetic make-up of the Ti plasmids. There are, however, still vast areas of the maps to which no functions have as yet been ascribed. On a molecular and biochemical level, we know very little concerning the functions of any of the genes shown in Figs 3 and 4. Nor do we have any evidence that provides a clear justification for the apparent biological requirement of the extremely large sizes of the Ti plasmids. The answers to some of these questions will probably be soon forthcoming.

III. THE FUTURE OF CROWN GALL RESEARCH

Although it may seem presumptuous to predict the future of any branch of science, there are a few concepts and features unique to the crown gall disease that allow one to say with some certainty that future research interest will concentrate in those areas. Practically speaking, the research areas listed below do not differ significantly from the areas of concentration of the past, but the list is helpful in that it points out the great diversity of scientific fields included in the disease.

A. Model System of Host–Pathogen Interactions at Many Levels

Crown gall is first and foremost a plant disease, and the study of the interaction between the pathogen and its many types of host plants can be performed at many biological levels.

1. Organismal

The study of bacterial-plant cell adherence may tell us much about the mechanisms and specificities of the adherence reaction. The effects, if any, that the pathogen and host have on each other during the early stages of wound healing and the nature of the wounding requirement are important questions to be answered.

2. Genetic

The genetic dissections of the processes involved in pathogenicity are well underway and future studies of the role that the chromosomal genes play in conferring oncogenicity to the bacterium should enhance our understanding of the disease.

3. Molecular

The elucidation at the molecular and biochemical levels of the exact processes that confer oncogenicity to the bacterium, and transform the plant cell, is much needed.

B. Model System for the Study of Plant Cell Growth and Regulation

Although some may validly argue that information gleaned in the study of abnormal cells has limited usefulness in describing normal cells, the dramatic physiologic aberrations characteristic of the crown gall tumour cells provide a system wherein processes thought to control normal plant cell growth are deficient or exaggerated. The study of the control, or loss of control, in the tumour cells, and the study of processes involved in regeneration of plant structures from tumour cells, may provide clues to the operation of those processes in the normally developing cell.

C. Model System of Cancer Studies

Crown galls are, by all conventional definitions, cancer neoplasias and are unique in their bacterial etiology. No other bacterially-induced cancers of plant or animal, have been described. And yet the Ti plasmids and mammalian tumour viruses share a common etiological mechanism in the integration of foreign DNA of causal agent origin into the host cell genome. An intriguing question concerns the possible existence of cancer-inducing bacteria having animal hosts. The evolutionary origin of the tumour-inducing Ti plasmid genes is also of concern, since Thomashow *et al.* (1980a) reported base sequence homology between Ti plasmid DNA and DNA isolated from a normal plant. This correlation brings to mind the presence of homologous sequences between the retroviruses and their non-infected hosts (Todaro and Huebner, 1972; Aaronson and Stephenson, 1976).

A further feature which renders the crown gall cancers amenable to future research is the process of suppression of the tumour traits, and regeneration of apparently normal cells. Clearly these topics are of timely importance.

D. Model System of Microbial Evolution

The evolution of genes conferring oncogenicity to a bacterium seems remarkable in itself, but the co-evolution of the complex biological processes surrounding the opines is truly astonishing. If the nopaline and octopine genes have arisen from separate ancestral catabolic plasmids, then the possibility exists that the phenomenon of genetic parasitism is not limited to the agrobacteria. Similar reactions may occur between *Rhizobium* species, which are closely related to the agrobacteria, and the plants which they nodulate. Other biotypes of agrobacteria may be discovered that will bridge the gap between *Agrobacterium* and *Rhizobium*. On an ecological basis, the observation that most isolates of agrobacteria from soil are nonvirulent (L. Moore, personal communication) may be of importance. Obviously, there are many questions yet to be answered.

E. Model System for the Study of Prokaryotic Gene Expression in a Eukaryotic Cell

Apart from the genetic engineering considerations to be discussed below, crown gall provides a unique system in which genes that normally reside in a prokaryotic cell are expressed naturally in a eukaryotic cell. Or, perhaps, the genes for oncogenicity and opine production are remnants of ancestral eukaryotic genes that now reside primarily in a prokaryotic cell. In either case, the mechanisms of control of the expression of those genes in their respective genetic backgrounds present intriguing problems. It is not known if the RNAs transcribed from T-DNA are processed beyond the additions of caps and poly-A tails (e.g. removal of intervening sequences), if they are translated using nonmodified host components, or if unique transfer RNA molecules are required. The analogies of the T-DNA to many eukaryotic virus infections suggest many possible perturbations in the host synthetic apparatus that may be altered upon infection.

F. Potential Vector for Plant Genetic Engineering

One of the most exciting aspects of the crown gall system concerns the potential use of the Ti plasmid, or portions of it, for the introduction of foreign DNA into plant cells. The rationale for these experiments is that since the opine synthesis genes are non-essential for tumour induction, they can be considered as sequences carried along into the plant cells during transfer of the common DNA sequences to those cells. Accordingly, one might expect that perhaps any DNA substituted into the Ti plasmid at the position of the opine synthesis genes would be transferred to the plant cell. That this seems to be the case was indicated by evidence presented by Hernalsteens *et al.* (1980). In their experiments, the bacterial transposon Tn7 was first integrated into the *Nos* gene of pTiC58. Upon examination of tumour DNA isolated from a crown gall incited by bacteria harbouring this mutant Ti plasmid, it was discovered that the tumour DNA did contain sequences homologous to the Tn7. Although no further characterization of this DNA as regards its stability and expression has been published, we can take hope from the fact that at least these first results were not negative.

The genetic engineering potential of the Ti plasmids remains unproven to date. Some questions which remain to be answered include the following.

(i) Will the insertion of an engineered gene at any position on the plasmid (except within the common DNA) be sufficient to allow its transfer to plant cells, or must the insertion be in the T-DNA?

(ii) Must the engineered gene bear its own promoter, or is read-through possible in the plant from adjacent plasmid or plant promoters?

(iii) Will mutant Ti plasmids be found which allow transfer of DNA to the plant cell without concomitant tumour induction?
(iv) Is the whole Ti plasmid necessary for engineering purposes, or can a shortened vehicle be constructed?
(v) Is it necessary that the engineering gene be integrated into the Ti plasmid, or can the Ti plasmid mobilize other plasmids present in the bacterial cell?
(vi) Are viable bacteria necessary, or can a plant cell transformation system using naked DNA be developed?
(vii) If such a transformation system is developed, will this enable the engineering of monocotyledons, plants that are not affected by the crown gall disease?
(viii) What is the stability of the engineered genes? If they are lost during meiosis, does that feature limit the engineering potential of the Ti plasmid to plants that are clonally (vegetatively) propagated?

The answers to some of these questions must be acquired before experiments designed to answer other questions can begin. At any rate, the possible use of the Ti plasmid in the creation of new plant varieties bearing gene combinations never before possible is certainly the most exciting aspect of future crown gall research. I feel there will be enough practical applications for such a genetic engineering vehicle that the efforts devoted to its construction will be well rewarded. The development of new crop varieties through genetic engineering techniques will probably never surpass the accomplishments of classical plant breeders, but it may make their job easier through the ability to move blocks of genes from different sources into a genetic background as a single unit, rather than through a time-consuming series of genetic crosses.

IV. Factors Contributing to Elucidation of the Crown Gall Mechanism

It is often said that the biggest obstacle to research is lack of adequate funding, and to a certain extent this is true. There is no doubt that crown gall science would be several years behind its present state were it not for the generous support of the National Institutes of Health, the National Science Foundation, the American Cancer Society, and the various European and Australian agencies that have funded crown gall researchers. But funding is only part of the answer, and in the author's opinion, it is the training of the researchers engaged in the study of crown gall that, more than in any other plant disease system, has contributed to the rapid progress of the last few years. This is not to say that all crown gall scientists are supremely gifted, but rather to point out that the backgrounds of many researchers comprise areas of training outside a rigid plant pathology programme. Someone once succinctly stated that "it was the microbiologists, not the plant pathologists, who finally solved crown gall". The point that the person (a microbiologist) was making is that, at the time that we stopped studying the disease from the plant's point of view, and turned instead to studies of the micro-organism involved, we encountered a major change in the philosophy of studying the disease. Admittedly, the studies of tumour physiology and the kindred studies of tumour induction provided information concerning the disease, but those studies have contributed very little to understanding the molecular basis of the disease. The reason for this should be obvious. One simply cannot perform the appropriate genetic experiments with the plant host that will reveal anything about the mechanism of the disease. By studying the bacterium, and mutants of the bacterium, we are able to perturb the system in ways that allow us to acquire a bit of information concerning the role of the mutated gene in inducing the disease. We moved then from a description of phenomenology to one wherein the scientist, and not the system, is in control of the experiment.

Not every disease needs to be understood at the same level as crown gall. Some would probably say that we already know too much about crown gall, and, from a plant pathology point of view, this may be so. The nurseryman does not need to know the structure of agrocin 84 to know that his biocontrol measures with strain 84 are working. On the other hand, there are scores of plant diseases about which we know virtually nothing. I consider it remarkable that, except for the studies of Fulbright and Leary (1978) with *Pseudomonas glycinea*, and the maps of *Erwinia amylovora* and *E. chrysanthemi* discussed by Chatterjee (1978), there are no other bacterial plant pathogens for which even a modest chromosomal gene map has been construced. It is an open question whether or not a thorough knowledge of other diseases on the molecular genetical level would result in better control of those diseases, but certainly by ignoring studies of the disease at that level, we will never know. It is my firm belief that the answers to many of plant pathology's questions could be gained through a thorough study of the genetics of the micro-organisms involved.

To assure adequate training for this type of research, early in a student's career the decision should be made as to which area of plant pathology, applied or basic research, the student wishes to pursue. If the interest is in basic research, then the student's training should include as many courses in biochemistry, cell biology, molecular biology, and genetics as possible, even if it means omitting some applied plant pathology courses. If crown gall is to be any example at all, let it be that it demonstrates the degree to which modern researchers, an interdisciplinary approach, and the availability and implementation of new scientific techniques, have been combined to provide an atmosphere for rapid research progress which has been unequalled in the study of any other plant disease.

V. ACKNOWLEDGEMENTS

I acknowledge the research support of the United States Department of Agriculture during the preparation of the manuscript. I thank Ann Fitzpatrick for assistance with the bibliography and artwork, and Christy Cox for typing the manuscript.

REFERENCES

Aaronson, S. A. and Stephenson, J. R. (1976). *Biochim. Biophys. Acta* **458,** 323–354.

Anderson, A. R. and Moore, L. W. (1979). *Phytopathology* **69,** 320–323.

Beimann, K., Lioret, C., Asselineau, K., Lederer, E. and Polonsky, J. (1960). *Bull. Soc. Chim. Biol.* **42,** 979–991.

Bomhoff, G., Klapwijk, P. M., Kester, C. M. and Schilperoort, R. A. (1976). *Mol. Gen. Genet.* **145,** 177–181.

Braun, A. C. (1947). *Amer. J. Bot.* **34,** 234–240.

Braun, A. C. (1958). *Proc. Natl. Acad. Sci. USA* **44,** 344–349.

Braun, A. C. (1978). *Biochim. Biophys. Acta* **516,** 167–191.

Braun, A. C. and Wood, H. N. (1976). *Proc. Natl. Acad. Sci. USA* **73,** 496–500.

Calos, M. P. and Miller, J. H. (1980). *Cell* **20,** 579–595.

Cavara, F. (1897). *Stn. Sper. Agrar. Ital. Modena* **30,** 482–509.

Chatterjee, A. K. (1978). *Proc. 4th Inter. Conf. Plant Path. Bact.* (Angers, France). pp. 3–16.

Chilton, M.-D., Drummond, M. H., Merlo, D. J. and Sciaky, D. (1978). *Nature* (London) **275,** 147–149.

Chilton, M.-D., Drummond, M. H., Merlo, D. J., Sciaky, D., Montoya, A. L., Gordon, M. P. and Nester, E. W. (1977). *Cell* **11,** 263–271.

Chilton, M.-D., Saiki, R. K., Yadav, N. and Gordon, M. P. (1980). *Proc. Natl. Acad. Sci. USA* **77,** 4060–4064.

Coxon, D. T., Davies, A. M. C., Fenwick, G. R. and Self, R. (1980). *Tetrahedron Lett.* **21,** 495–498.

Currier, T. C. and Nester, E. W. (1976). *J. Bacteriol.* **126,** 157–165.

Davis, C. H. and Rothman, R. H. (1973). *Mut. Res.* **20,** 283–285.

Depicker, A., Van Montagu, M. and Schell, J. (1978). *Nature* (London) **275,** 150–153.

Drummon, M. (1979). *Nature* (London) **281,** 343–347.

Drummond, M. and Chilton, M.-D. (1978). *J. Bacteriol.* **136,** 1178–1183.

Drummon, M. H., Gordon, M. P., Nester, E. W. and Chilton, M.-D. (1977). *Nature* (London) **269,** 535–536.

Einset, J. W. (1980). *Plant Physiol.* **65,** S24.

Ellis, J., Kerr, A., Tempé, J. and Petit, A. (1979). *Mol. Gen. Genet.* **173,** 263–269.

Engler, G., Holsters, M., Van Montagu, M. and Schell, J. (1975). *Mol. Gen. Genet.* **138,** 345–349.

Firmin, J. L. and Fenwick, G. R. (1977). *Phytochemistry* **16,** 761–762.

Firmin, J. L. and Fenwick, G. R. (1978). *Nature* (London) **276,** 842–844.

Fulbright, D. W. and Leary, J. V. (1978). *J. Bacteriol.* **136,** 497–500.

Gantotti, B. V., Patil, S. S. and Mandel, M. (1978). *Proc. 3rd Inter. Congr. Plant Pathol.* (Munich, Fed. Rep. Ger.). p. 68.

Genetello, C. (1978). *Proc. 4th Inter. Conf. Plant Path. Bact.* (Angers, France). p. 262.

Genetello, C., Van Larebeke, N., Holsters, M., Depicker, A., Van Montagu, M. and Schell, J. (1977). *Nature* (London) **265,** 561–563.

Goldmann, A., Tempé, J. and Morel, G. (1969). *C.R. Hebd. Seances Acad. Sci.*, Ser. B **268,** 852–854.

Gonzalez, C. F. and Vidaver, A. K. (1979). *Curr. Microbiol.* **2,** 75–80.

Gresshoff, P. M., Skotnicki, M. L. and Rolfe, B. G. (1979). *J. Bacteriol.* **137,** 1020–1021.

Gurley, W. B., Kemp, J. D., Albert, M. J., Sutton, D. W. and Callis, J. (1979). *Proc. Natl. Acad. Sci. USA* **76,** 2828–2832.

Guyon, P., Chilton, M.-D., Petit, A. and Tempé, J. (1980). *Proc. Natl. Acad. Sci. USA* **77,** 2693–2697.

Hack, E. and Kemp, J. D. (1977). *Biochem. Biophys. Res. Commun.* **78,** 785–791.

Hamada, S. E., Luckey, J. P. and Farrand, S. K. (1979). *J. Bacteriol.* **139,** 280–286.

Hamilton, R. H. and Fall, M. Z. (1971). *Experientia* **27,** 229–230.

Hepburn, A. G. and Hindley, J. (1979). *Mol. Gen. Genet.* **169,** 163–172.

Hernalsteens, J.-P., Van Vliet, F., De Beuckeleer, M., Depicker, A., Engler, G., Lemmers, M., Holsters, M., Van Montagu, M. and Schell, J. (1980). *Nature* (London) **287,** 654–656.

Holsters, M., de Waele, D., Depicker, A., Messens, E., Van Montagu, M. and Schell, J. (1978). *Mol. Gen. Genet.* **163,** 181–187.

Holsters, M., Silva, B., Van Vliet, F., Genetello, C., De Block, M., Dhaese, P., Depicker, A., Inzé, D., Engler, G., Villarroel, R., Van Montagu, M. and Schell, J. (1980). *Plasmid* **3,** 212–230.

Depicker, A., Inzé, D., Engler, G., Villarroel, R., Van Montagu, M. and Schell, J. (1980). *Plasmid* **3,** 212–230.

Jensen, C. O. (1910). 2e Conf. Int. pour l'Étude du Cancer. Arsskr. K. Vet.-Landbohoejsk., Copenhagen. pp. 243–254.

Jensen, C. O. (1918). Arsskr. K. Vet.-Landbohoejsk., Copenhagen. pp. 91–143.

Kado, C. I. (1976). *Annu. Rev. Phytopathol.*, 265–308.

Keane, P. J., Kerr, A. and New, P. B. (1970). *Aust. J. Biol. Sci.* **23,** 585–595.

Kelly, T. J. and Nathans, D. (1977). *Adv. Virus Res.* **21,** 86–173.

Kemp, J. D. (1977). *Biochem. Biophys. Res. Commun.* **74,** 862–868.

Kemp, J. D. (1978). *Plant Physiol.* **62,** 26–30.

Kennedy, B. W. and Alcorn, S. M. (1980). *Plant Dis.* **64,** 674–676.

Kerr, A. (1969). *Nature* (London) **223,** 1175–1176.

Kerr, A. and Htay, K. (1974). *Physiol. Plant Pathol.* **4,** 37–44.

Kerr, A. and Panagopoulos, C. G. (1977). *Phytopath. Z.* **90,** 172–179.

Kerr, A., Manigault, P. and Tempé, J. (1977). *Nature* (London) **265,** 160–161.

Kersters, K., Deley, J., Sneath, P. H. A. and Sackin, M. (1973). *J. Gen. Microbiol.* **78,** 227–239.

Koekman, B. P., Ooms, G., Klapwijk, P. M. and Schilperoort, R. A. (1979). *Plasmid* **2,** 347–357.

Ledeboer, A. M. (1978). "Large Plasmids in *Rhizobiaceae*". Dissertation (University of Leiden, The Netherlands).

Limsmaier, E. M. and Skong, F. (1965). *Physiol. Plant.* **18,** 100–127.

Lippincott, J. A., Webb, J. H. and Lippincott, B. B. (1965). *J. Bacteriol.* **90,** 1155–1156.

Lippincott, B. B. and Lippincott, J. A. (1969). *J. Bacteriol.* **97,** 620–628.

Lippincott, J. A. and Lippincott, B. B. (1975). *Annu. Rev. Micro.* **29,** 377–405.

Liv, S.-T. and Kado, C. I. (1979). *Biochem. Biophys. Res. Commun.* **90,** 171–178.

Loper, J. E. and Kado, C. I. (1979). *J. Bacteriol.* **139,** 591–596.

Marton, L., Wullems, G. J., Molendijk, L. and Schilperoort, R. A. (1979). *Nature* (London) **277,** 129–131.

Matthysse, A. G. and Stump, A. J. (1976). *J. Gen. Microbiol.* **95,** 9–16.

Matthysse, A. G., Wyman, P. M. and Holmes, K. V. (1978). *Infect. Immun.* **22,** 516–522.

McPherson, J. C., Nester, E. W. and Gordon, M. P. (1980). *Proc. Natl. Acad. Sci. USA* **77,** 2666–2670.

Ménagé, A. and Morel, G. (1964). *C.R. Hébd. Séances Acad. Sci.*, Ser. B **259,** 4795–4796.

Ménagé, A. and Morel, G. (1965). *C.R. Hébd. Séances Acad. Sci.*, Ser. B **261,** 2001–2002.

Merlo, D. J. (1975). "Sunflower crown Gall: A Molecular Approach to the Mechanism of Tumor Induction", Dissertation (University of Wisconsin, Madison).

Merlo, D. J., Nutter, R. C., Montoya, A. L., Garfinkel, D. J., Drummon, M. H., Chilton, M.-D., Gordon, M. P. and Nester, E. W. (1980). *Mol. Gen. Genet.* **177,** 637–643.

Messens, E. and Claeys, M. (1978). *Proc. 4th Inter. Conf. Plant Path. Bact.* (Angers, France). pp. 169–170.

Milani, V. and Heverlein, G. T. (1972). *J. Virol.* **10,** 17–22.

Montoya, A. L., Chilton, M.-D., Gordon, M. P., Sciaky, D. and Nester, E. W. (1977). *J. Bacteriol.* **129,** 101–107.

Montoya, A. L., Moore, L. W., Gordon, M. P., and Nester, E. W. (1978). *J. Bacteriol.* **136,** 909–915.

Moore, L. W. (1977). *Phytopathology* **67,** 139–144.

Moore, L. W. and Schroth, M. N. (1976). *Phytopathology* **66,** 1460–1465.

Murphy, P. J. and Roberts, W. P. (1979). *J. Gen. Microbiol.* **114,** 207–213.

New, P. B., Kerr, A. (1972). *J. Appl. Bacteriol.* **35,** 279–287.

Panagopoulos, C. G., Psallidas, P. G. and Alivizatos, A. S. (1978). *Proc. 4th Inter. Conf. Plant Path. Bact.* (Angers, France). pp. 221–228.

Petit, A. and Morel, G. (1966). *C.R. Séances Soc. Biol. Ses Fil.* **160,** 1806.

Petit, A., Delhaye, S., Tempé, J. and Morel, G. (1970). *Physiol. Veg.* **8,** 205–213.

Roberts, W. P., Tate, M. E. and Kerr, A. (1977). *Nature* (London), **265,** 379–381.

Rogler, C. E. (1980). *Proc. Natl. Acad. Sci. USA* **77,** 2688–2692.

Roizes, G., Pages, M. and Lecou, C. (1979). *Gene* **6,** 43–50.

Sacristan, M. D. and Melchers, G. (1977). *Mol. Gen. Genet.* **152,** 111–117.

Saedler, H., Reif, H., Hu, S. and Davidson, N. (1974). *Mol. Gen. Genet.* **132,** 265–289.

Schell, J., Van Montagu, M., De Beuckeleer, M., De Block, M., Depicker, A., De Wilde, M., Engler, G., Genetello, C., Hernalsteens, J. P., Holsters, M., Jeurinck, J., Silva, B., Van Vliet, F. and Villarroel, R. (1979) *Proc. R. Soc. Lond. B.* **204,** 251–266.

Schroth, M. N. and Hildebrand, D. C. (1968). *Phytopathology* **58,** 848–854.

Sciaky, D., Montoya, A. L. and Chilton, M.-D. (1978). *Plasmid* **1,** 238–253.

Smith, V. A. and Hindley, J. (1978). *Nature* (London) **276,** 498–500.

Smith, E. F. and Townsend, C. O. (1970). *Science* **25,** 671–673.

Sonoki, S. and Kado, C. I. (1978). *Proc. Natl. Acad. Sci. USA* **75,** 3796–3800.

Tempé, J., Petit, A., Holsters, M., Van Montagu, M. and Schell, J. (1977). *Proc. Natl. Acad. Sci. USA* **74,** 2848–2849.

Thomashow, M. F., Nutter, R., Montoya, A. L., Gordon, M. P. and Nester, E. W. (1980a). *Cell* **19,** 729–739.

Thomashow, M. F., Panagopoulos, C. G., Gordon, M. P. and Nester, E. W. (1980b). *Nature* (London **283,** 749–796.

Todaro, C. J. and Huebner, R. J. (1972). *Proc. Natl. Acad. Sci. USA* **69,** 1009–1015.

Turgeon, R., Wood, H. N. and Braun, A. C. (1976). *Proc. Natl. Acad. Sci. USA* **73,** 3562–3564.

Upper, C. D., Helgeson, J. P., Kemp, J. D. and Schmidt, C. J. (1970). *Plant Physiol.* **45,** 543–547.

Van Larebeke, N., Genetello, C. H., Schell, J., Schilperoort, R. A., Hermans, A. K., Hernalsteens, J. P. and Van Montagu, M. (1975). *Nature* (London) **255,** 742–743.

Watson, B., Currier, T., Gordon, M. P., Chilton, M.-D. and Nester, E. W. (1975). *J. Bacteriol.* **123,** 255–264.

Whatley, M. H., Margot, J. B., Schell, J., Lippincott, B. B. and Lippincott, J. A. (1978). *J. Gen. Microbiol.* **107,** 395–398.

White, P. R. and Braun, A. C. (1941). *Science* **94,** 239–241.

Wilson, T., Papahadjopoulos, D. and Taber, R. (1979). *Cell* **17,** 77–84.

Wood, H. N., Binns, A. N. and Braun, A. C. (1978). *Differentiation* **11,** 175–180.

Yang, F., McPherson, J. C., Gordon, M. P. and Nester, E. W. (1980a). *Biochem. Biophys. Res. Commun.* **92,** 1273–1277.

Yang, F., Montoya, A. L., Merlo, D. J., Drummond, M. H., Chilton, M.-D., Nester, E. W. and Gordon, M. P. (1980b). *Mol. Gen. Genet.* **177,** 707–714.

Zaenen, I., Van Larebeke, N., Teuchy, H., Van Montagu, M. and Schell, J. (1974). *J. Mol. Biol.* **86,** 109–127.

Zambryski, P., Holsters, M., Kruger, K., Depicker, A., Schell, J., Van Montagu, M. and Goodman, H. (1980). *Science* **209,** 1385–1391.

5

ENZYMES INVOLVED IN THE REPLICATION OF PLANT VIRAL RNAs

T. C. Hall[1], W. A. Miller[2] and J. J. Bujarski[2]

[1] *Agrigenetics Advanced Research Laboratory, Madison*
[2] *Department of Horticulture, University of Wisconsin, Madison*

I. Abbreviations
II. Concepts: The Not-so-simple Virus
III. Definition of Replicase
IV. RNA Polymerase Activity in Healthy Plants
V. Critical Evaluation of Views Questioning the Existence of Virally-coded Replicase
VI. Putative Structure of Viral Replicase in Plants
VII. Replicases of Bacterial RNA Viruses
VIII. Replicases of RNA Viruses of Animals
IX. Replicases from Plant Hosts
  A. Introduction: Strategies for Isolation
  B. Brome Mosaic Virus Replicase
  C. Cowpea Chlorotic Mottle Virus Replicase
  D. Cowpea Mosaic Virus Replicase
  E. Tobacco Mosaic Virus Replicase
  F. Alfalfa Mosaic Virus Replicase
  G. Turnip Yellow Mosaic Virus Replicase
  H. Pea Enation Mosaic Virus Replicase
  I. Cucumber Mosaic Virus Replicase
X. Future Directions

ABBREVIATIONS

cDNA	copy DNA
ds	double stranded
EF	elongation factor
EGTA	ethyleneglycol-*bis*-(β-aminoethyl ether)N,N′-tetraacetic acid
kd	kilodalton
mRNA	messenger RNA
PAGE	polyacrylamide gel electrophoresis
RF	replicative form
RI	replicative intermediate
RNase	ribonuclease
ss	single stranded

	Detergents:
12-M	dodecyl-β-D-maltoside
NP-40	Nonidet-P40
	Viruses:
AMV	alfalfa mosaic virus
BBMV	broad bean mosaic virus
BMV	brome mosaic virus
CMV	cucumber mosaic virus
CCMV	cowpea chlorotic mottle virus
CPMV	cowpea mosaic virus
EMCV	encephalomyocarditis virus
EMV	eggplant mosaic virus
PEMV	pea enation mosaic virus
STNV	satellite tobacco necrosis virus
TMV	tobacco mosaic virus
TYMV	turnip yellow mosaic virus
TNV	tobacco necrosis virus

II. CONCEPTS: THE NOT-SO-SIMPLE VIRUS

Viruses are often cited as simple forms of life, having but few biological functions. Indeed, it is clear that their genetic information content *per se* must be less than that of organisms capable of independent existence. However, there is much evidence to show that viruses are very highly adapted, using many succinct ruses to enable their small amount of nucleic acid to encode multiple functions.

A relatively straightforward adaptation is that of overlapping genes, initially established for the DNA virus $\phi\chi174$, where certain regions are used in two or all three reading frames (Barrell *et al.*, 1976). RNA viruses also have overlapping genes; the lysis protein of MS12 overlaps the regions coding for coat and replicase (Beremand and Blumenthal, 1979; Atkins *et al.*, 1979). Plant viruses also have overlapping genes, examples being TMV and TYMV (Davies, 1979).

Influenza is an orthomyxovirus, the protein-coding information of which is present in encapsidated (−) RNA strands which have to be transcribed to give (+) strands capable of serving as mRNAs. Such viruses encapsidate functional polymerase. Although (−) RNAs synthesized from plant viral RNAs have not been shown to encode proteins, this may simply reflect the small quantities of such strands typically available for study. Matthews (1978) noted that a relatively long overlapping gene can be derived from the complementary sequence of potato spindle tuber viroid.

A list of subtleties used by viruses must also include the possibility that many of the multiple bands often seen in electrophoretic profiles of cell-free translation products coded by plant virus RNAs represent functional proteins. These bands are usually reproducible for a given virus, and optimization of reaction conditions to yield "full length" products may, in fact, represent a more artifactual state than that obtained using less finely tuned systems. Several pieces of evidence can be given in support of this statement. Pelham (1978) and Shih *et al.* (1978, 1979) have shown that cell-free translation of encephalomyocarditis (EMCV) and poliovirus RNAs yields new proteolytic activities that are involved in the processing of other products of cell-free translations. Similar observations have been obtained with CPMV RNA translation products (Pelham, 1978).

Van Tol and van Vloten-Doting (1979) have shown that alfalfa mosaic virus RNA1 can be translated to yield a polypeptide of 115 kd or two polypeptides of 58 and 62 kd, depending upon RNA concentration. Supplementation of the reaction with tryptophan results in enhanced synthesis of the longer polypeptides, suggesting that the tryptophan codon functions in some controlling manner (van Vloten-Doting *et al.*, 1980). Although no biological function has yet been assigned to any of these three molecules, a contribution to processes regulating infection cannot be excluded.

Processing of viral proteins was initially shown for the "polyprotein" synthesized from poliovirus RNA (Röder and Koschel, 1975). A series of processing events that lead to an active replicase for EMCV has been suggested by Rueckert *et al.* (1980), and this is discussed in more detail below (Section VIII).

The interaction of viral proteins with host protein is another ruse that effectively multiplies the information content of the viral genome. The classic example is $Q\beta$ replicase. Of the four polypeptides that combine to form active and specific enzyme, only one protein is virally coded (see section VII).

It is probable that other viral–host protein or protein–RNA (or DNA–DNA, and all combinations) interactions as well as modifications (glycosylation, methylation, hydroxylation, aminoacylation) contribute to what we believe to be the formidable information content of these "simple" pathogens. However, the examples cited above are adequate to help explain why so little is known about a fundamental viral process, that of replication of its genome.

III. DEFINITION OF REPLICASE

A polymerase is an enzyme that condenses individual nucleotides into a

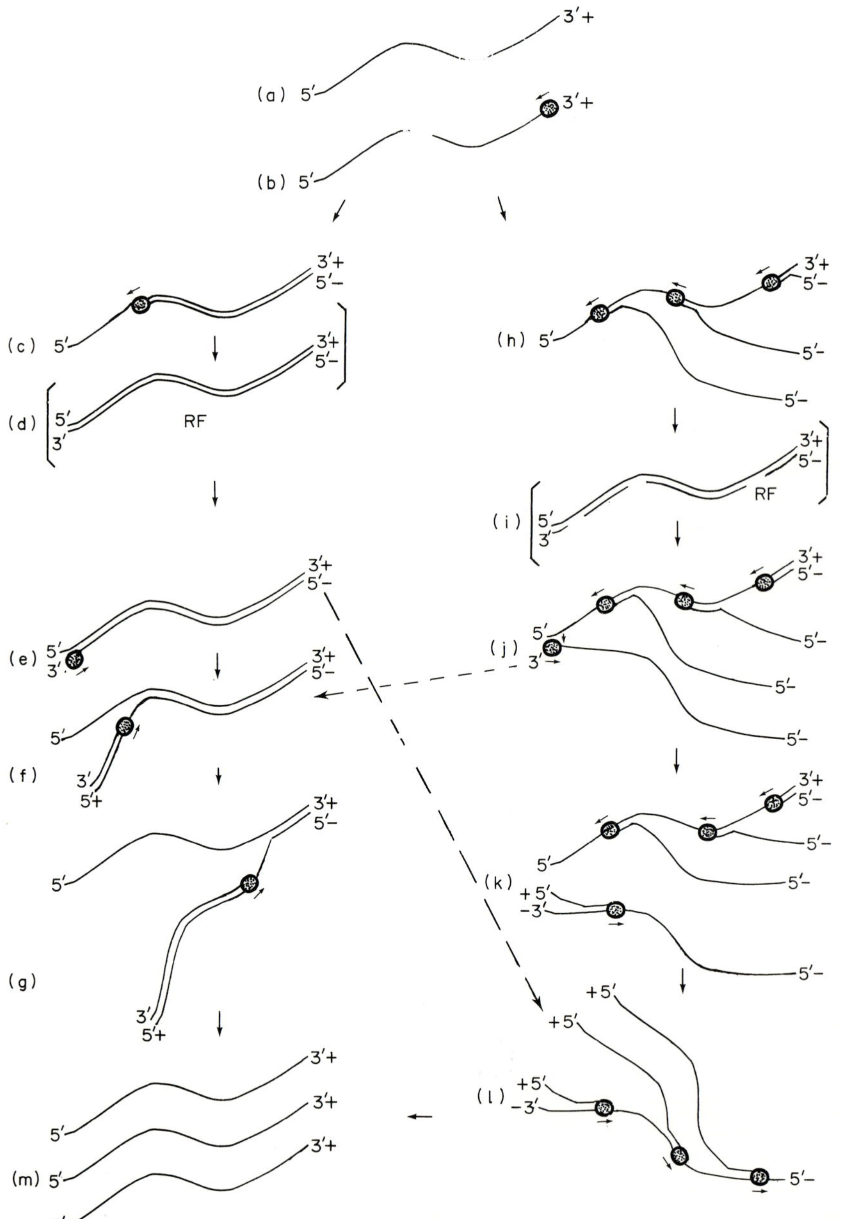
(a) 5′ 3′+
(b) 5′ 3′+
(c) 5′ 3′+ 5′−
(d) 5′ 3′ RF 3′+ 5′−
(e) 5′ 3′ 3′+ 5′−
(f) 5′ 3′ 5′+ 3′+ 5′−
(g) 5′ 3′ 5′+ 3′+ 5′−
(m) 5′ 5′ 5′ 3′+ 3′+ 3′+
(h) 5′ 3′+ 5′− 5′− 5′−
(i) 5′ 3′ RF 3′+ 5′−
(j) 5′ 3′ 3′+ 5′− 5′− 5′− 5′−
(k) +5′ −3′ 5′ 3′+ 5′− 5′− 5′− 5′−
(l) +5′ −3′ +5′ +5′ 5′−

Fig. 1.

polynucleotide chain. A template may not be necessary, e.g. for the polyadenylating enzyme that attaches often lengthy tracts of adenosines to the 3′ terminus of mRNAs, both cellular and viral. However, most polymerases do use a template which can be DNA or RNA. The various forms of DNA-dependent RNA polymerases have been reviewed in detail (Chambon, 1975; Roeder, 1976; Becker, 1979) and will not be further mentioned here; neither will RNA-dependent DNA polymerase (Temin and Baltimore, 1972), also known as reverse transcriptase. A true replicase must be able to make not only the initial complementary RNA copy (typically the (−) strand) of the RNA template, but also to use the (−) strand to make a new second (typically (+) in the sense that it serves as a messenger for protein synthesis) strand that is a replica of the initial RNA template. Replication starts at the 3′ end of the template RNA, although the actual binding site of the replicase to the template is unlikely to be at the 3′ extremity itself. A partially completed copy is called a replicative intermediate and is often represented diagrammatically as shown in Fig. 1. Full-length dsRNA, when isolated free of protein, is designated as the replicative form (RF).

The RF shown in Fig. 1d implies that a single polymerase attaches to a template whereas the RF in Fig. li results from several polmerase molecules travelling along the template at a given time, the single-stranded tails being lost as a result of nuclease cleavage or during the isolation procedure. Analytically, it should be simple to distinguish between the covalent RF of Fig. 1d and the "nicked" RF shown in Fig. li by electrophoresis on denaturing gels; newly synthesized radioactive (−) strand should be intact if the pathway represented by a→b→c→d is correct. As will be described later, few *in vitro* replication systems described thus far have sufficient activity to permit useful resolution of the products. Replicative form isolated from intact plants appears to be

---

Fig. 1. Possible pathways of viral RNA replication. First, (−) strand RNA is made from the (+) strand template (a→d or a→b→h→i) to yield the RF (d or i) that would be obtained upon extraction of double stranded RNA. Next, (+) strand RNA is made from the (−) template (e→g→m or j→m). In either stage of replication, one RNA polymerase molecule may replicate one RNA molecule (left-hand column), or many polymerase molecules may each replicate a single RNA (right-hand column). Hybrid schemes such as a→d→e→l→m or a→h→j→f→g→m can be envisaged. Indeed, it is quite likely that the former is the actual sequence because this would result in the synthesis of far more (+) than (−) strand, for translation and encapsidation while minimizing energy expended in synthesizing multiple (−) strand intermediates. See text for further discussion.

of the structure shown in Fig. 1d, but this may be artifactual in representing the reassociation of (−) and (+) forms to give a relatively nuclease-resistant double stranded form rather than reflecting what occurs during replication (Garnier *et al.*, 1980). Synthesis of the second strand is pictured in Fig. 1e→g→m and j→m. We believe the sequence 1 e→l→m to more closely represent the actual situation because of the very small amount of (−) RNA typically present in virally-infected plant tissues.

IV. RNA POLYMERASE ACTIVITY IN HEALTHY PLANTS

In healthy tobacco plants, enzymatic activity has been found which can: (i) incorporate radioactive nucleotide triphosphates into an acid precipitable form, (ii) synthesize RNase-resistant products, (iii) synthesize small, heterodisperse RNA products, (iv) synthesize RNA that appears to be complementary to the RNA offered as template, (v) synthesize RNA without any template specificity (Duda *et al.*, 1973; Duda, 1979; Romaine and Zaitlin, 1978; Clerx and Bol, 1978; Chifflot *et al.*, 1980). In healthy cowpeas, enzyme fractions with similar properties have been reported (White and Dawson, 1978), but complementarity of the products to template has not been established. Enzyme preparations that can incorporate radioactive UTP into trichloroacetic acid-precipitable material have been reported in chinese cabbage (Astier-Manifacier and Cornuet, 1971), in tobacco and cowpea (Ikegami and Franekel-Conrat, 1978b; Zabel *et al.*, 1981) and in tomato (Boege and Sänger, 1980). In the latter case, tomato was reported to have ten times the activity found in tobacco, using the procedure of Ikegami and Fraenkel-Conrat (1978b). Boege and Sänger emphasize the importance of ruling out nucleotidyl transferase activities as do Zabel *et al.* (1981), who show that a terminal uridylyl transferase activity is greatly increased upon infection of cowpeas with CPMV. Unfortunately, none of these papers shows any analysis of products for size or complementarity.

In all cases where it has been studied, this RNA-dependent RNA polymerase activity is stimulated by traumatization of the plant. It appears likely that it is also greatly stimulated on viral infection. This has led to the conjecture that viral replication is carried out entirely by the host plant, and that virus infection only stimulates the plant's RNA-dependent RNA polymerase activity (Fraenkel-Conrat, 1979). However, no such plant polymerase has been shown to fulfil the definition of a replicase, i.e. to use (+) stranded viral RNA as a template to make many more faithful copies of full length (+) stranded RNA.

RNA-dependent RNA polymerase in healthy plants has been suggested as being responsible for the replication of viroids which are thought to lack the capacity to code for their own replicase (Boege and Sänger, 1980). However, Sänger's group recently found that PSTV appears to use host-supplied DNA-dependent RNA polymerase II as its replicase, although the activity of this enzyme using viroid RNA as template is rather low (Rackwitz *et al.*, 1981).

V. CRITICAL EVALUATION OF VIEWS QUESTIONING THE EXISTENCE OF VIRALLY-CODED REPLICASE

Strong statements made by Ikegami and Fraenkel-Conrat (1980) supported the surmise that the weak activity present in healthy plants is the one used to replicate viruses. Because of the significance of this concept, it is useful to analyse the evidence leading to the belief that viral replicases are host-encoded. Fraenkel-Conrat (1979) has summarized such views. Firstly, it is claimed that no detectable differences in properties existed for enzyme preparations from healthy, TMV-infected, or TNV-infected tobacco. Unfortunately, the activities of the enzyme fractions cited (Ikegami and Fraenkel-Conrat, 1978a) were very low, being less than 400 cpm [^{3}H]UMP incorporated into acid-insoluble materials. The data do not permit a meaningful assessment of activity (e.g. in terms of picomoles UMP incorporated per unit of enzyme). In only a few experiments was there any stimulation (less than two-fold) over the endogenous reaction when RNA was added; in many, no significant stimulation was obtained. Such data do not provide convincing evidence for a lack of differences in properties of the "enzyme" preparations.

Romaine and Zaitlin (1978) also reported no qualitative differences in crude polymerase fractions from uninfected and TMV-infected tobacco. The activity of the enzyme preparations was low, and no template specificity was observed although hybridization data were used to support the claim that both enzyme preparations yielded sequences complementary to TMV RNA. It would have been instructive to have included a control experiment with the mock-inoculated enzyme in which no TMV RNA was added as template in order to determine if any product that would anneal to TMV RNA was synthesized. An additional control would be to add other viral RNAs; production of complementary RNAs would indicate lack of template specificity and consequently cast doubt on the involvement of that enzymatic activity in TMV RNA replication. In line with such concepts, these authors note that their enzyme is "not the complete replicase" and entertain the possibility that a virally-coded polypeptide might associate with host-coded proteins to form a "fully-competent replicase".

In contrast to the above data, Hardy *et al.*, (1979) showed major differences in enzyme fractions isolated from healthy, mock-inoculated and BMV-infected barley leaves. This work, and more recent studies, are discussed below, but suffice it to say here that the alleged lack of differences for "enzymes" from healthy and virally-infected plant tissues rests on very weak evidence.

A second aspect used by Fraenkel-Conrat (1979) to support the notion that replication is due to host-encoded enzymes is the observation that enzyme preparations from different plants infected with the same virus are different. Again, the data (Ikegami and Fraenkel-Conrat, 1978b) are for very crude preparations having low activity, showing little or no stimulation by added RNA templates, and for which no product data were obtained. It is hardly surprising that crude enzyme preparations from different plants show different activities in terms of counts per minute incorporated.

The third aspect raised by Fraenkel-Conrat (1979) in support of the theory that there is no difference between polymerase from uninfected and infected plants is that viral infection causes the enzyme activity to increase in proportion to the increase in total protein in their extract. However, in their cogent paper Ikegami and Fraenkel-Conrat (1979) admit that they provide only circumstantial evidence.

A fourth suggestion (Fraenkel-Conrat, 1979; Ikegami and Fraenkel-Conrat, 1978a) is that the inability of most laboratories to isolate a template specific enzyme implies that the enzyme is host-supplied. The problem here is that specificity often appears to be lost during purification, especially during solubilization from membranes and in attempts to free the enzyme of endogenous template (see Section VI).

The data relating to template utilization (Fraenkel-Conrat, 1979) involve many subjective evaluations of preparations having low activity. It is not clear from the data presented if they refer to healthy or virus-infected tissues. The arguments regarding template specificity are similar to those used concerning the lack of differences in enzymes from healthy and infected material.

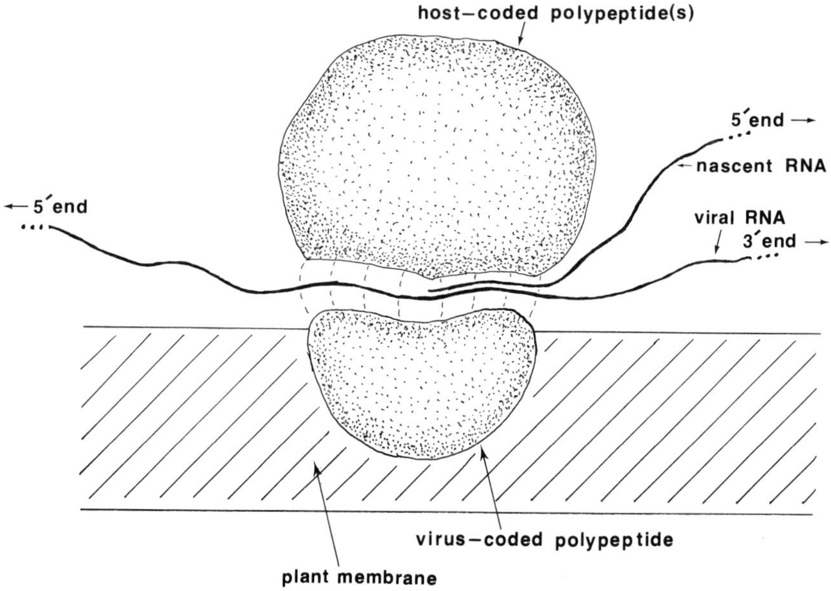

Fig. 2. Hypothetical model of the RNA-dependent RNA polymerase replication complex. The host-coded polypeptide(s) has non-specific polymerase activity and can readily be removed from the membrane, and the RNA. The virus-coded polypeptide is embedded in the membrane and confers specificity for the viral RNA to the replication complex. It has a strong affinity for the, as yet unspecified, plant membrane and for viral RNA.

VI. PUTATIVE STRUCTURE OF VIRAL REPLICASE IN PLANTS

Data to be discussed below for BMV RNA replicase strongly support the concept that a virally coded enzyme does indeed exist. However, the ease with which template specificity can be lost during purification procedures, especially those involving removal of endogenous RNA, are in accord with the notion that it is a complex enzyme. There have been several reports that the RNA-dependent RNA polymerase activity in healthy plants can be found in the soluble fraction whereas the much greater virus-induced activity is associated with membranes. The loss of template-specific activity noted when any of several detergents are used to solubilize the enzyme can be interpreted as reflecting an association of the putative virus specific (presumably virus-encoded) peptide with the plant membrane. A summary of the available evidence gives rise to the model in Fig. 2. This model can be used to explain the confusion

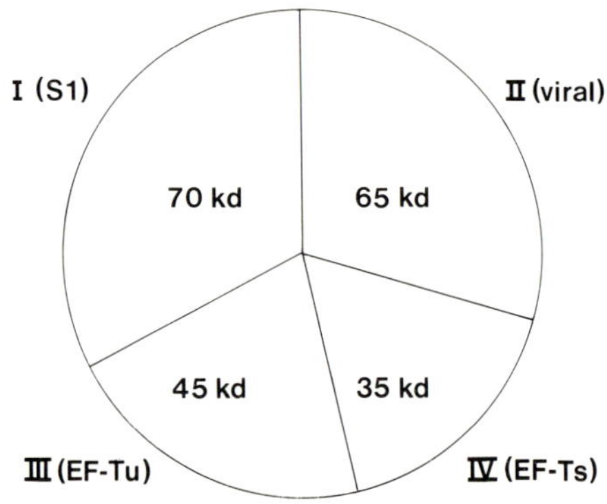

Fig. 3. Diagrammatic representation of the Qβ replicase complex showing relative contributions of the subunits. An additional protein (host factor) is also known to be involved in the replication of Qβ RNA (Blumenthal and Carmichael, 1978).

surrounding the similarities between the activities seen on solubilization of enzyme from infected plants and from uninfected tissues. In both cases the researchers are dealing only with the host-encoded part of the replicase.

It is of course a moot point as to whether all viral replicases are similar in structure, being composed of host and virally coded polypeptides. Indeed, they may well not be. Although there seems to be little reason why the replicase could not be entirely virally coded, the best defined systems all show involvement of host factors. As methodology for handling membrane bound (or membrane associated) enzymes improves, we have little doubt that the participation of a virally coded polypeptide will be seen as an essential feature of viral replication.

VII. REPLICASES OF BACTERIAL RNA VIRUSES

Replication of bacterial viruses was intensively studied during the 1970s and a valuable review was written by Blumenthal and Carmichael (1979). Therefore, only the aspects most relevant to a discussion of plant virus replicases will be summarized here. The definitive work of Weissman *et al.* (1963) showed that Qβ replicase was highly specific for its own RNA, as the RNA of the closely related phage MS2 was poorly recognized except in conditions specially developed to relax template specificity (e.g. the addition of Mn^{2+} to the reaction; Palmenberg and Kaesberg, 1974). The work of Blumenthal and Landers (1976) and subsequent studies established that only one (65 kd) of the four subunits comprising Qβ replicase (Fig. 3) is virally coded. Surprisingly, on a molecular weight basis, one-third of the complex consists of protein synthesis elongation factors EF-Tu and EF-Ts. The ribosomal polypeptide S_1 appears to be essential for binding and initiation of RNA synthesis on $Q\beta$ (+) strands, but not (−) strands. A host factor present in uninfected as well as in infected cells appears to be involved in the synthesis of (−) strands directed by the synthesis of $Q\beta$ (+) strands. Host factor is thought to be a heat stable oligomer of six identical 12 kd chains associated with ribosomes but of unknown function in healthy cells (Carmichael *et al.*, 1975).

VIII. REPLICASES OF RNA VIRUSES OF ANIMALS

The RNA-dependent RNA polymerases (replicases) of RNA viruses which infect animal cells were discovered about 20 years ago (Baltimore and Franklin, 1962; Weissman *et al.*, 1963). Attempts to isolate specific replicases from animal tissues infected with RNA viruses have met with many of the troubles encountered in plant systems.

Most information available relates to the replication of picornaviruses. Baltimore's group studying poliovirus multiplication isolated an active polyuridylating enzyme (Flanegan *et al.*, 1977). They obtained fractions that could elongate RNA chains but which were incapable of initiating new RNA strands. Further studies on purification and protein composition of picornavirus replicase (Traub *et al.*, 1976; Rueckert *et al.*, 1980) as well as on the processing of proteins (Palmenberg, 1979; Pelham, 1978) allowed formulation of a possible mechanism for the generation of viral replicase. A scheme for the processing of individual EMCV-coded proteins as well as their role in EMCV multiplication is explained in Fig. 4. In general, this scheme reveals a way in which endoproteolytic activity connected with replicase can serve in regulation of replication, translation and encapsidation, the three main processes involved in multiplication of viruses.

IX. REPLICASES FROM PLANT HOSTS

A. Introduction: Strategies for Isolation

We have already indicated that the study of plant virus replicase systems is far from easy. Because the phenotypic appearance of infection symptoms is easily seen and because biochemical documentation of viral RNA and protein accumulation can readily be obtained over defined periods following inoculation of a healthy plant, it would seem that the

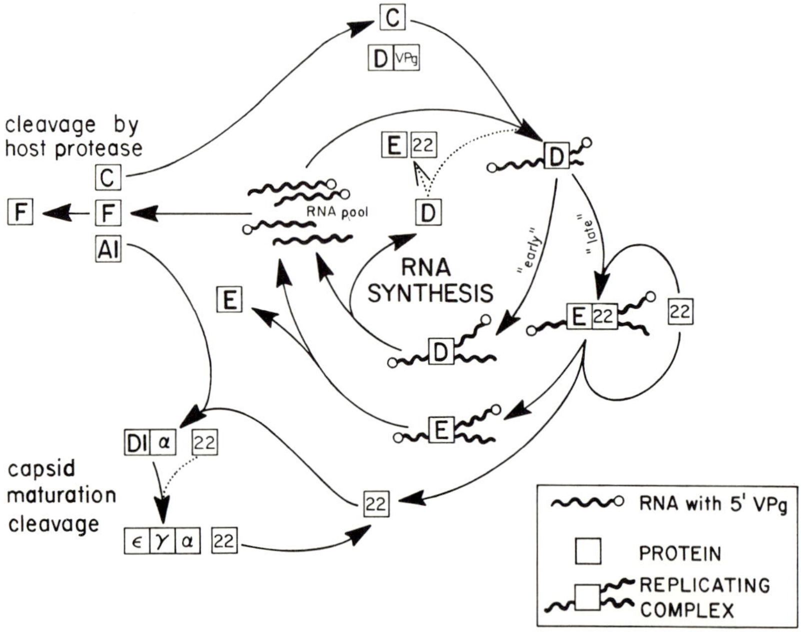

Fig. 4. A hypothetical scheme proposed by Dr A. Palmenberg for the processing of individual EMCV-coded proteins. EMCV-encoded precursor protein C consists of three parts, each having a different role during multiplication. The peptide fragments are released in a stepwise manner during processing, mediated by proteolytic activity of p22. Autoproteolytic precursor C is encoded at the 3′-end of picornaviral RNA. It releases VPg protein (Pallansch *et al.*, 1980) which immediately connects with the 5′-end of the nascent RNA strand, rendering it capable of initiating replication. Replication is carried out further by protein D early in the infection, or by protein E at late stages of infection. The cleavage of D to E and p22 releases free p22 protease fragments which probably participate in the processing (cleavage) of precursor A into capsid proteins ε, γ and α. Protein E is incapable of reinitiating new viral strands so that the concentration of mRNA decreases, influencing the translation process.

enzyme systems involved should readily be isolated. Indeed, some of the work in the early 1970s showed that addition of nucleotide triphosphates *in vitro* to crude extracts of infected plants resulted in their polymerization into acid-precipitable forms. Using gel-slicing techniques Semal and Kummert (1970) provided evidence that BMV RNAs were synthesized in infected barley tissue. However, little progress towards purification was made and the insensitive product assay technique available at that time left much to be desired.

The studies of Zabel *et al.* (1974, 1976, 1979); Semal and Kummert (1970), Hardy *et al.* (1979) and several others have accentuated the need for collecting the plant material at a critical stage of development of infection. This requirement may well reflect the existence of a processing step, or steps, similar to those of EMCV polymerase (Fig. 4) in the generation of active replicase. The active form of the enzyme might, therefore, be prevalent only when vigorous synthesis of molecules processed to the initiation-active (IA) form has occurred and very little subsequent processing to the elongation-active but initiation-inactive (EA) form has taken place. The IA form would thus be capable of initiation and elongation, analogous to the D polypeptide of EMCV replicase. As infection progresses, the EA form accumulates and substantially decreases the proportion of the IA form. As separation procedures improve, the isolation of the IA form at all stages of infection where replication is taking place can be anticipated.

Thus far, we have stressed the need to recognize the existence of host RNA-dependent RNA polymerase in healthy plants and to differentiate the stimulation of activity of this enzyme on traumatization or infection from the existence of true viral replicase. The host enzyme is template non-specific (or, at least not viral RNA-specific) whereas the viral enzyme should be template-specific. This precaution has proven to be especially cogent in the case of CPMV replicase since Dorssers *et al.*, (1982) have shown that the reports of Zabel *et al.* (1974, 1976, 1979) actually related to a host-encoded polymerase, whose involvement in viral replication is unknown, rather than the enzyme functional in viral replication.

Other important aspects involve the reaction conditions. Ethanol is a poison to replication and the fact that the radioactive nucleotide triphosphate precursors are frequently supplied in ethanol has undoubtedly caused much grief. It is also important to recognize that GTP is highly unstable, and when delivered or stored at temperatures above $-60°C$, it rapidly loses potency.

A further complication for RNA polymerase reactions *in vitro* is that they have a fairly high Km for UTP (e.g. 15·3 μM for BMV; Hardy *et al.*, 1979). This fact was recognized by Symons (1977) and Kumarasamy and Symms (1979). Unfortunately, it is impractical to add radioactive precursors to reactions at this level and dilution of specific activity results in lowered detectability of products. A method to circumvent this difficulty is to initiate reactions at low UTP concentration (high specific radioactivity) and then augment chain completion by adding non-radioactive UTP.

For certain steps in isolation we have found high ionic conditions to be detrimental and the presence of membrane fragments to be important. Because membranes contain many proteins, the presence of membranous material in polymerase preparations is a major source of impurity. The attachment of polymerase to membranes accounts for their being found in the "particulate" fractions of crude extracts. The procedures used to release polymerase appear to be only partially effective in that they decrease the size (or density) of the enzyme-membrane complex. As a result, the enzyme is solubilized in the sense that it is not sedimented by centrifugation at a given force over a short period (e.g. 2–20 000 *g* for 10–20 min). However, centrifugation at high forces (e.g. 100 000 *g*) for even short times will sediment the replication-active enzyme.

Investigators have used many procedures to obtain solubilization of plant replicases by detergent treatment (Triton X-100, lubrol-W, SDS) or by Mg^{2+}-depletion approaches. We have observed that such treatments do solubilize activity, but consistently result in the loss of template specificity. A conclusion consistent with such observations is that a viral-specific moiety of the enzyme is embedded in the membrane (see Fig. 2). This conclusion has led us to use dodecyl-β-D maltoside (12-M) as detergent during extraction. This detergent appears to serve as a functional membrane-substitute. Its use has resulted in the isolation of highly active template-specific and template-dependent enzyme that is stable for several months when frozen (Bujarski *et al.*, 1982).

To summarize, replicase isolation probably requires the establishment of vigorous, close to synchronous, infection, the use of non-ionic extraction conditions and meticulous attention to reaction component quality and quantity.

B. Brome Mosaic Virus Replicase*

Early publications on BMV replicase by Semal and co-workers (Semal and Hamilton, 1968; Semal and Kummert, 1970; 1971) established that 10 000 *g* particulate fraction extracts from BMV-infected barley leaves incorporated much more [^{3}H]UMP into acid-precipitable product than did similar extracts from healthy leaves. The development of actinomycin-D-resistant activity was detected two days after inoculation, reached a peak three days after inoculation, and declined over the next three days. Most of the products were shown to be RNase-resistant, but pulse-chase experiments provided evidence that label incorporated during a 3–5 min pulse into RNase-resistant product could be shifted into RNase-sensitive product during the 10–15 min chase (Semal and Kummert, 1971). These data suggested that initially double-stranded

---

* In these sections, the term replicase is used not because preparations capable of *de novo* (+) strand synthesis have been attained, but rather because that is the ultimate objective of the studies.

[(+) strand template together with new (−) strand] material gave rise to new (+) strands. Further characterization of products (Kummert and Semal, 1974; Semal and Kummert, 1976) and partial solubilization of the polymerase with NP-40 (Kummert and Semal, 1977) leave little doubt that a virally-induced enzyme was present in infected leaves that catalysed [^{3}H]UMP incorporation into replicative forms of BMV RNAs. The data of Kummert and Semal (1977) are highly suggestive that the activity is BMV RNA-specific, but these workers did not clearly show initiation onto the supplied template and the need for further purification of the extract was recognized (Semal and Kummert, 1976). Hadidi and Fraenkel-Conrat (1973) obtained a BMV RNA-stimulatable enzyme from BMV-infected barley by solubilization with Triton X-100. Some preferences for BMV RNA template over other viral RNAs (TMV, STNV, BBMV) was observed. However, little discrimination against CCMV RNA was obtained.

Initial attempts in our laboratory to isolate viral replicase from BMV-infected (and healthy) barley leaves yielded data similar to that published by other groups, i.e. [^{3}H]UMP incorporation levels of 2–10 000 cpm and little stimulation by the addition of BMV RNA template. Higher activities (about 2-fold) were generally obtained for extracts from inoculated leaves or from traumatized leaves than from healthy ones. Attempts to purify the enzyme via DEAE-cellulose or DEAE-Sephadex chromatography using procedures analogous to those of Zabel *et al.* (1974, 1976) were fruitless. Removal of endogenous template by the Mg^{2+}-depletion techniques of Zabel *et al.* (1974, 1976) was also unsuccessful. Furthermore, the enzyme activity was lost on freezing. Enzyme stability is of great practical significance. It is difficult to maintain stocks of suitable plant material, tedious to repeat extractions on essentially a daily basis, and impossible to undertake comparative enzymological experiments.

After elimination of high salt conditions, removal of ethanol from the radioactive precursors and attention to the quality of supplied GTP, striking differences were obtained between extracts from healthy and infected plants. These included a high total UMP incorporating activity for the extracts, initially 10-fold better than the previous data and subsequently 100- to 1000-fold better. The activity of extracts from BMV-infected plants was often 70-fold higher than that from mock-inoculated plants (Hardy *et al.*, 1979), and excellent template dependence and specificity (Table I) were apparent. Until one has obtained such differences, we believe it is hard to appreciate the distinction between the host polymerase and viral replicase. Undoubtedly, this is reflected in papers stating that viral replication depends on a host enzyme.

Table I. Characteristics of BMV replicase treated with micrococcal nuclease.

Template Added	$[^{32}P]$UMP incorporated (cpm) Untreated	Nuclease Treated
None (0 min)	4 165	3 297
None (30 min)	71 585	5 962
CCMV RNA (30 min)	84 332	14 155
BMV RNA (30 min)	207 740	133 230
Template Dependence[a]	67%	98%
Template Specificity[b]	91%	94%

Replicase was prepared as described by Hardy *et al.* (1979) except that 12-M was used instead of NP-40 as detergent, and the final enzyme pellet was obtained by centrifugation at 100 000 *g* instead of being precipitated with polyethylene glycol. Reactions (60μl) included replicase (40 μl) in assay buffer (50 mM Tris pH 8·0, 10 mM $Mg(OAc)_2$,) and 0·4 mM $Ca(OAc)_2$. Nuclease treatment (Pelham and Jackson, 1976) was achieved by addition of 2·4 μg micrococcal nuclease (Sigma), and incubation for 15 min at 20°. The digestion was terminated by adding EGTA (pH 8·0) to a final concentration of 16·7 mM. Nuclease-treated, or untreated controls were assayed for replicase activity by the addition of 0 or 15 μg BMV or CCMV RNA (5 μl) to the reactions, 1 μl actinomycin D (500 μg ml^{-1}) and 100 pmol $[^{32}P]$UTP (specific activity 100 Ci $mmol^{-1}$). After incubation for the indicated periods at 30°, 40 μl samples were spotted on trichloroacetic acid-saturated discs, washed and counted as described by Hardy *et al.* (1979). In separate experiments, it was shown that neither $Ca(OAc)_2$ nor EGTA affected replicase activity, and that micrococcal nuclease, in the absence of EGTA, eliminated $[^{32}p]$UMP incorporation.

a: Template dependence = [(cpm with BMV RNA) − (cpm without added template)] × 100/[(cpm with BMV RNA) − (cpm at 0 min)].

b: Template specificity = [(cpm with BMV RNA) − (cpm with CCMV RNA)] × 100/[(cpm with BMV RNA) − (cpm without added template)].

For the studies reported by Hardy *et al.* (1979), small amounts of tissue (1–5 g) were used as starting material, and frozen extracts treated with NP-40 retained activity for up to two weeks. Initiation of replication was demonstrated by the addition of individual BMV RNA components as templates (Fig. 5a). This approach was adopted because the high amounts of endogenous viral RNA in the replicase preparation effectively diluted the specific activity of radioactive template added in experiments designed to show protection from S1 nuclease on double-strand synthesis. Recently, we have eliminated endogenous viral RNA without altering the template specificity or other properties of the enzyme using micrococcal nuclease (Table I) in a procedure modified from that of Pelham and Jackson (1976). This further verifies the initiation of replication by this enzyme and shows that endogenous RNA is not needed to obtain high template specificity.

Dodecyl-β-D-maltoside (12-M) appears to be very useful in enhancing stability of BMV polymerase, and also in purification steps to remove contaminating RNA and protein. Figure 6 shows the polypeptide composition of a highly active BMV RNA-dependent and BMV RNA-specific enzyme from infected barley purified by sucrose density gradient centrifugation in the presence of 12-M. This procedure increased specific activity some 200 times, yielding an enzyme which was stable on freezing for several months. The presence in this extract of a polypeptide that co-migrates with, and has the same tryptic peptide map as, the *in vitro* translation product of BMV RNA 1 suggests that this virally-coded polypeptide contributes to replicase activity. This band was not present in similar enzyme preparations from mock-inoculated tissues (Fig. 6). Evidence from work with barley protoplasts (Loesch-Fries and Hall, 1980) showed that BMV RNAs 1 or 2 encode a polypeptide or polypeptides involved in replication (Kiberstis *et al.*, 1981). As shown in Fig. 5b, double-stranded BMV RNAs are obtained *in vitro* using this partially-purified enzyme.

^{32}P-labelled RNA products from such extracts hybridized to cDNA synthesized using BMV RNA as template (Fig. 7). This provides preliminary evidence for *de novo* synthesis of (+) strand RNA. The possibility that (−) strand ^{32}P-RNA had hybridized to cDNA is doubtful because a control experiment showed that ^{32}P-labelled cDNA did not hybridize to unlabelled cDNA bound to nitrocellulose under identical conditions. Whether the (+) strand ^{32}P-RNA replicase products are synthesized using endogenous (−) strand RNA or newly synthesized (−) strand as template remains to be determined.

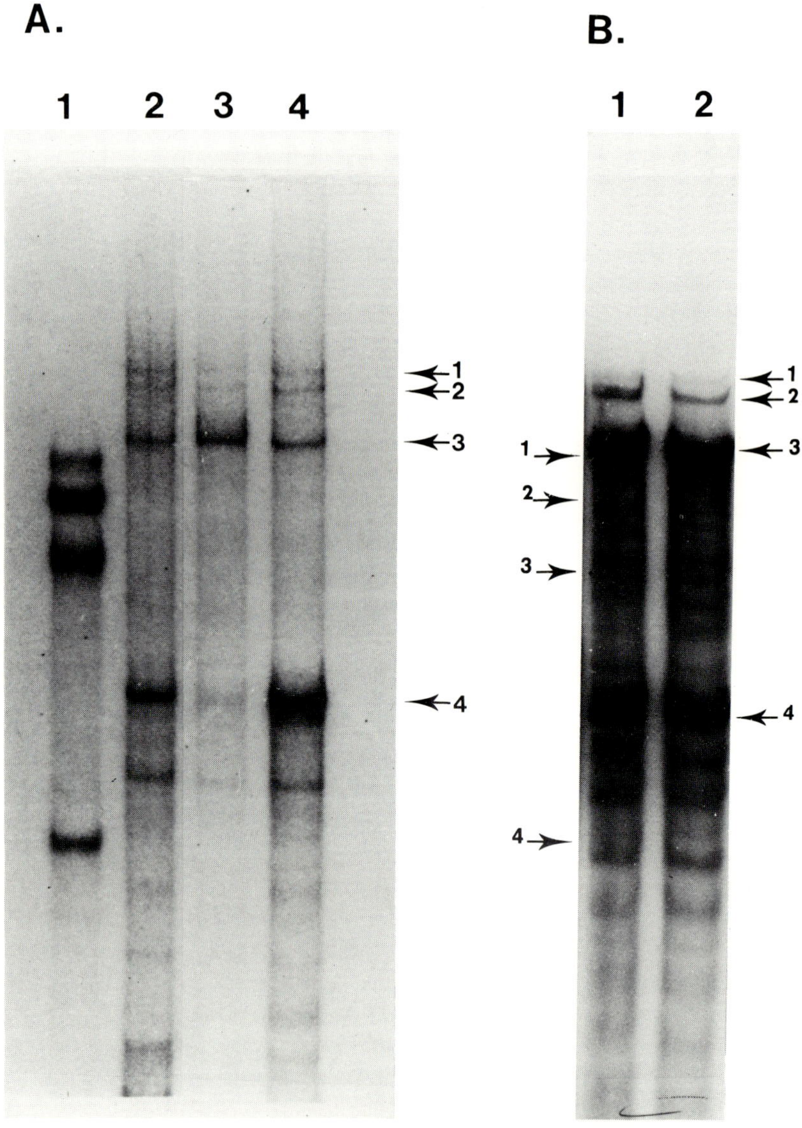
A.
1
2
3
4
1
2
3
4
B.
1
2
1
2
3
4
1
2
3
4

C. Cowpea Chlorotic Mottle Virus Replicase

A membrane-bound replicase from CCMV-infected cowpeas has been reported by White and Dawson (1978). This crude extract, which was not solubilized, showed neither RNA dependence nor template specificity. The highest level of activity was only 1896 cpm of [^{3}H]UMP incorporated; however, the specific activity of the isotope was not given. Using gel-slicing techniques, they found their products to be small and heterodisperse.

An extract prepared from CCMV-infected cowpea leaves (W. A. Miller, unpublished observations) by the method of Bujarski *et al* (1982) readily incorporated [^{3}H]UMP into acid-precipitable material (~ 150 000 cpm or 7–8 pmoles), but the reaction was not template-specific and fluorography of the products after polyacrylamide-agarose gel electrophoresis showed only very faint bands corresponding to full length CCMV RF RNAs; most of the radioactivity was in small, heterodisperse products. While the incorporation could have been misinterpreted as replicase activity, the use of fluorography revealed that most of this incorporation should in fact be attributed to a nonspecific host polymerase.

D. Cowpea Mosaic Virus Replicase

Zabel *et al.* (1974) isolated a particulate fraction from CPMV-infected cowpea leaves that incorporated [^{3}H]UMP into RNase-resistant and acid-precipitable products in actinomycin-D-insensitive reactions. Subsequently (Zabel *et al.*, 1976), solubilization was achieved by washing a 31 000 *g* pellet containing the bound enzyme with a Mg^{2+}-deficient buffer. This enzyme was stable on storage at 4°C and could be further

Fig. 5. Autoradiograph of 2·4% acrylamide—0·5% agarose gel electrophoresis of polynucleotide products synthesized by various preparations of BMV polymerase. Labelled nucleic acids were phenol-extracted from reaction mixtures containing [^{32}P]UTP as described by Hardy *et al.* (1979).

(A) Products of BMV polymerase were prepared (Hardy *et al.*, 1979) from reaction mixtures containing:unfractionated BMV RNA (lane 2); BMV RNA highly enriched in RNA 3 (lane 3); BMV RNA highly enriched in RNA 4 (lane 4). Lane 1 contains single stranded genomic RNA isolated from BMV-infected barley protoplasts. Arrows indicate mobilities of BMV replicative forms.

(B) Products from BMV polymerase prepared using 12-M treatment (Lane 1) and from polymerase pelleted by sucrose gradient centrifugation after 12-M treatment (Lane 2). Arrows on left indicate mobilities of ss BMV virion RNA, arrows on right indicate mobilities of BMV replicative forms.

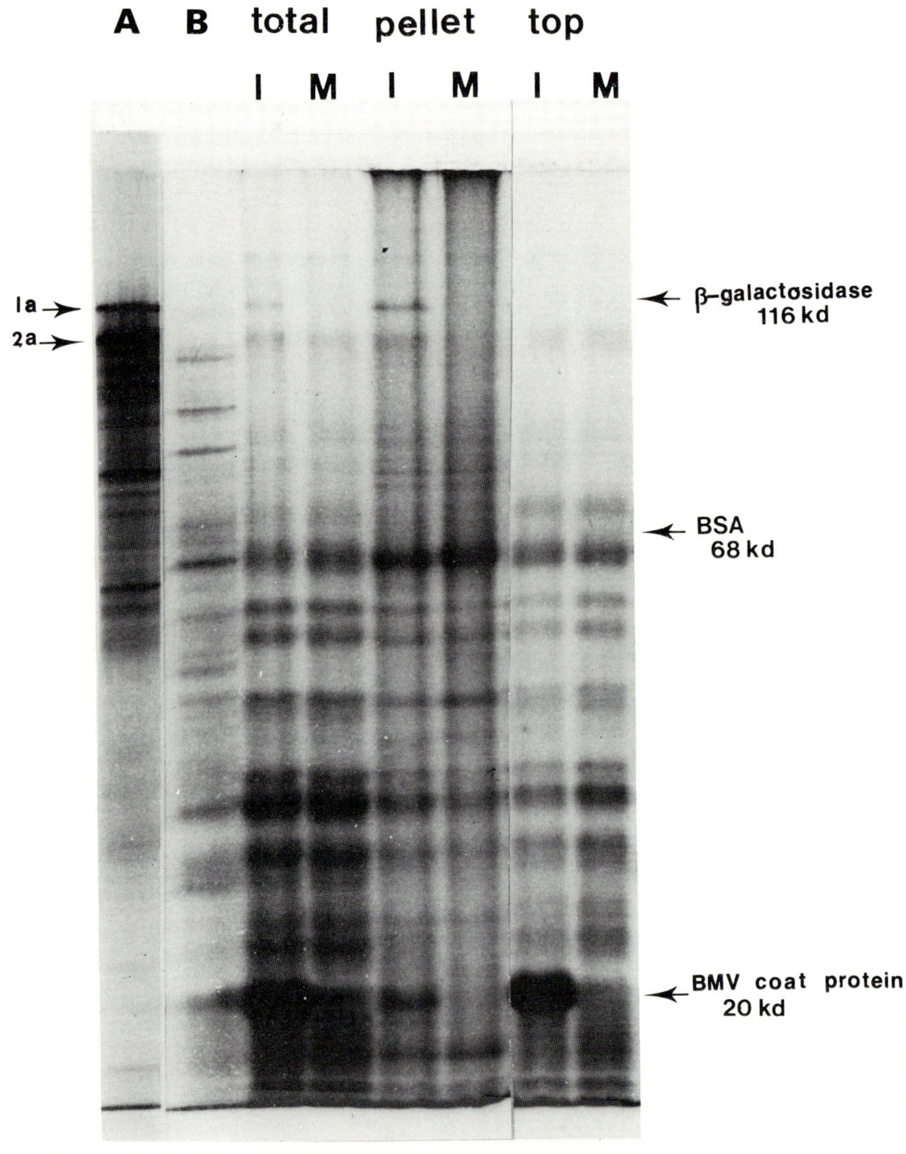

Fig. 6. SDS-polyacrylamide (16% w/v) gel electrophoresis of polymerase preparations (25 μg protein per well) from BMV-infected (I) and mock-inoculated (M) barley leaves. The lanes marked *total* represent the 12-M treated extract before gradient centrifugation; those marked *pellet* and *top* denote fractions collected after centrifugation. Lane A is an autoradiograph of wheat germ translation products of BMV RNAs 1 + 2; lane B shows the Coomassie-stained components present in the *in vitro* reaction. The positions of marker proteins are indicated together with their molecular weights at the right. Note the 110 kd band is present only in extracts with replicase activity, i.e. from infected tissue only. This band co-migrates with the translation product of BMV RNA 1.

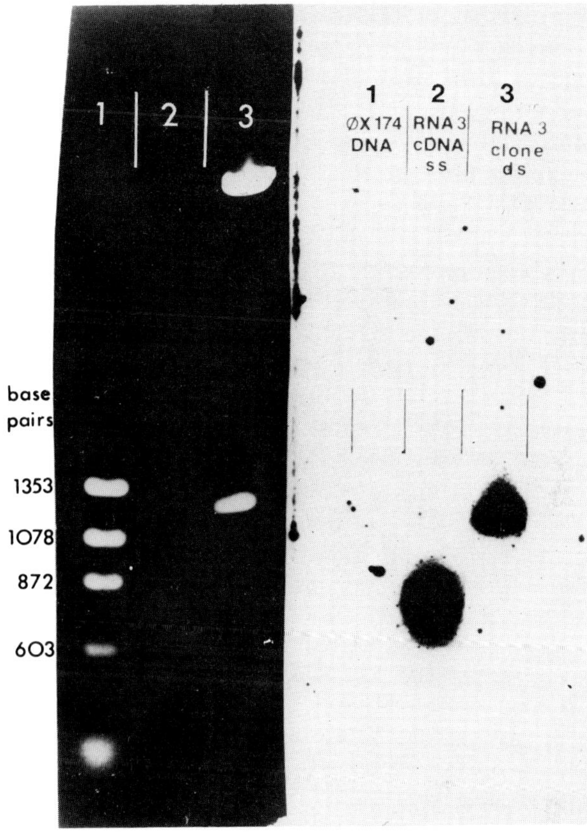

Fig 7. RNA-DNA Hybridization showing synthesis of (+) strand RNA by BMV polymerase. Single-stranded (ss) cDNA to the 3a cistron of BMV RNA 3 was synthesized using a procedure modified from that of Ahlquist *et al.* (1981). A nearly full-length double-stranded (ds) cDNA copy of this cistron was then synthesized and inserted into the *PstI* site of pBR322 by standard recombinant DNA methods (Weinand *et al.*, 1979; P. A. Kiberstis, unpublished data). Both the sscDNA and the dsDNA clone (the latter digested with *PstI* to release the BMV-specific insert) were electrophoresed through a 1% agarose gel and blotted onto a nitrocellulose filter (Southern, 1975). ^{32}P-labelled RNA products synthesized by BMV replicase were extracted as described by Hardy *et al.* (1979), passed over a Sephadex G-50 column, and denatured by boiling. The high molecular weight eluate (170 000 cpm) was hybridized to the nitrocellulose for 24 h at 42°C. The filter was then washed extensively, dried, and exposed to X-ray film.

(A) Ethidium bromide-stained gel (B) Autoradiograph of the Southern blot following hybridization to ^{32}P-labelled products of BMV replicase. Lane 1: *Hae*III digest of ϕX174 RF DNA. Sizes of fragments are indicated in base pairs.
Lane 2: 0·128 μg sscDNA synthesized from BMV RNA 3 (the ethidium bromide staining of the sscDNA was too faint to be visible on photograph). Lane 3: 0·105 μg of dsDNA specific for BMV RNA3. The stained band of high molecular weight in lane 3 is pBR322 DNA from which the insert was excised with *PstI*.

purified on DEAE-BioGel to yield a template-dependent fraction that carried out [^{3}H]UMP incorporation at a linear rate for 9 h. Additional experiments (Zabel *et al.*, 1979) resulted in RNA synthesis that continued for at least 15 h. Although template specificity could not be demonstrated directly, the use of an assay that depended on the ability of nitrocellulose filters to bind nucleoprotein complexes, but not RNA itself, suggested preferential binding of the enzyme preparation to comovirus RNAs. While this indirect assay may reveal affinity of an enzyme preparation to bind specific RNAs, it does not demonstrate that the enzyme selectively uses this template in replication. Indeed, more recent work from the same laboratory (L. Dorssers *et al.* personal communication) has made it clear that the purification steps achieved by Zabel (1979), relate to the host-encoded RNA-dependent RNA polymerase. An electrophoretically homogeneous enzyme preparation from infected cowpea leaves was found to be identical to a similar preparation from mock-inoculated leaves. Notably absent were polypeptides coded by RNA of B-component which has been shown to be involved in CPMV replication (Goldbach, *et al.*, 1980). Current work (L. Dorssers *et al.* personal communication) shows that crude preparations that do contain B-component-encoded polypeptides from CPMV-infected cowpea leaves are capable of active incorporation of radioactive precursors into replicative intermediates and RF's of CPMV RNAs.

Zabel (1978) has discussed the possibility that template specificity may not be a requisite for CPMV replicase to function properly in the replication process. Replication of CPMV has been shown to take place in cytopathic structures characteristic of virus infected cells (Assink *et al.*, 1973; de Zoeten *et al.*, 1974). The suggestion follows that localization of replication may replace the need for template specificity. Although such arguments can be extended to suggest that viral replicases may simply have to discriminate against cellular RNAs rather than specifically select their own template, we believe that the best evidence available supports the notion of strict template specificity.

E. Tobacco Mosaic Virus Replicase

There are many reports on the isolation of TMV replicase from TMV-infected tobacco leaves, those by Heisenberg, and Schramm (1965), Ralph and Wojcik (1969), and Bradley and Zaitlin (1971) being early examples. Even among these early reports, confusion exists concerning whether the enzyme is soluble or in a bound state that requires the use of detergent for solubilization. Enzyme bound to small fragments of membrane may remain in the supernatant of low speed centrifugation treatments but can be pelleted by high speed centrifugation.

A soluble enzyme that was in the supernatant of a 1 h centrifugation at 100 000 *g* was described by Brishammer and Juntti (1974). The crude preparation contained poly(A) synthesizing activity which was partially removed by gel filtration on an 8% agarose gel (BioRad A-1·5). The enzyme was then further purified by affinity chromatography through Sepharose 2B coupled to yeast RNA. A small amount of material was obtained that incorporated about 24 000 cpm of [^{3}H]UMP (specific activity of supplied UTP, 15 Ci mmol^{-1}) in the standard 250 μl reaction used. This reaction was entirely dependent on added RNA for template and showed some template specificity. Although Qβ RNA, yeast RNA, and poly(U) were relatively inactive in stimulating UMP incorporation (3·4%, 4·1%, and 0·4% of the TMV RNA value, respectively), poliovirus RNA, STNV RNA, and clover yellow vein virus RNA gave stimulation of over 50% of the control values. These data suggest that much, if not all, of the putative viral-specific subunit had been lost during isolation. In this respect, it is interesting that the authors note that the apparent molecular weight of their enzyme (130 000 daltons; 6·5 S) is similar to that of Qβ replicase, and raise the question as to whether TMV replicase may be similarly constructed in having only one of its four subunits virally-encoded. The enzyme was very sensitive to freezing. Detailed product analysis was not undertaken (electrophoretic separation techniques were not well developed at that time), but the majority was shown to be double-stranded on the basis of RNase A and T$_1$-resistance under high salt conditions. Similar extracts from healthy tobacco leaves were reported to show no activity; apparently mock-inoculated leaves were not tested.

Romaine and Zaitlin (1978) isolated a soluble RNA-dependent RNA polymerase from TMV-infected tobacco leaves. Although no qualitative differences in polymerase activity of the enzyme from infected plants were detected in comparison with those for the enzyme from healthy leaves, the specific activity of the polymerase from infected material was apparently greater. Because a number of catalytic parameters for both enzymes were indistinguishable, these authors suggested that the enhancement of activity was due to a stimulation of a host polymerase activity following TMV infection. On the other hand, these authors did not exclude the possibility of association of some TMV-coded proteins with the host enzyme to form a fully competent and specific TMV replicase.

Sela and Hauschner (1975), using Celmusol as a detergent, obtained a TMV RNA-dependent polymerase activity from a 12 000*g* pellet fraction of TMV-infected tobacco leaves. Initial extracts from both infected and uninfected leaves showed no template specificity, but the extract from infected leaves became TMV-RNA specific after "ageing" the enzyme for 24 h. The results from this rather peculiar procedure are difficult to explain but they may reflect the different stabilities of individual components of a complex enzyme.

White and Murakishi (1977) investigated both soluble RNA-dependent RNA polymerase and membrane bound TMV replicase in tobacco callus inoculated with TMV. Initial extraction was in 150 mM Tris (pH 8·2) buffer containing 0·4 M sucrose, 10 mM KCl, 5 mM $MgCl_2$, 2·5 mM dithiothreitol and 10% glycerol (v/v). The soluble polymerase was isolated from 31 000*g* supernatant by lowering the pH to 4·7 with HCl. The enzyme was not stimulated with viral RNA templates and had properties identical to those of the enzyme from healthy callus. The crude, bound, enzyme also showed no template specificity. However, after solubilization by Mg^{2+}-depletion and elution with KCl from a DEAE-Sephadex column (Zabel *et al.*, 1976), fractions showing high template specificity (TYMV RNA was only 9% and BMV RNA only 21% as effective as TMV RNA) were obtained. This enzyme was calculated to have a molecular weight of 150 000 and was shown to incorporate [^{3}H]UMP into acid-insoluble material with linear kinetics for 6 h. Sucrose-SDS density gradient centrifugation of the RNA products yielded a peak of about 16S, believed to be the replicative form of TMV RNA. Considering the lengthy incorporation kinetics shown, it is surprising that no new (+) strand synthesis was detected. The enzyme was described as being "relatively stable". A similar fraction from healthy callus contained non-specific RNA polymerase activity.

Considering the significant properties claimed by White and Murakishi (1977) for their enzyme, it is noteworthy that Romaine and Zaitlin (1978) make only passing reference to their work. Unlike White and Murakishi (1977), Fraenkel-Conrat *et al.* assert that TMV exclusively uses non-specific host RNA-dependent RNA polymerase for replication of its RNA (Fraenkel-Conrat, 1979; Ikegami and Fraenkel-Conrat, 1980). These authors base their opinion on experiments which suggested that the enzymatic and physico-chemical properties of both soluble and solubilized membrane-bound enzymes isolated from different plants after infection with different viruses were similar to those from healthy plants. The activities of the preparations tested by Ikegami and Fraenkel-Conrat (1980) were generally very low. This, together with their failure to observe template-specific stimulation suggests to us that they failed to isolate the viral specific enzyme (see also comments in Section V, p. 185).

F. Alfalfa Mosaic Virus Replicase

Weening and Bol (1975) found that polymerase extracts of AlMV-infected broad beans exhibited differences from extracts of uninfected plants, a result that was not observed in similar extracts from tobacco. They state

> . . . that the products synthesized in the extracts of infected and healthy leaves are of a different nature becomes apparent when the phenol-extracted material, synthesized in the presence of actinomycin-D, is treated with RNase.

The overall activity of the Weening and Bol (1975) extracts is, in fact, not very high and on the basis of total acid-precipitable cpm of [^{3}H]UMP incorporated in the presence of actinomycin-D, there is but a two-fold difference between extracts from infected (7804 cpm) and healthy (3120 cpm) tissue. Reannealing experiments, which yielded only 1187 cpm of RNase-resistant material, were interpreted to show that "at least 92% of the label is in plus-strand AlMV". In the face of results presented by almost all other investigators showing the absence, or minimal synthesis, of new viral (+) strand RNA by crude extracts, reservations must exist as to the validity of this conclusion.

A later paper from the same laboratory (Clerx and Bol, 1978) further documents differences between extracts from healthy and AlMV-infected leaves (of tobacco plants) after solubilization of the enzyme by the Mg^{2+}-depletion procedure of Zabel *et al.* (1976). Little activity was obtained in comparable extracts from healthy leaves. This work is cited by Ikegami and Fraenkel-Conrat (1980) in support of their belief that healthy and virus-infected plants yield enzymes of similar properties. Such citation may be rationalized on the basis of statements made by Clerx and Bol (1978) who present a table showing incorporation of 8180 cpm by a Mg^{2+}-depletion-solubilized extract from infected leaves compared with 650 cpm from healthy leaves and conclude

> . . . the possibility must be considered that the virus-induced, soluble RNA polymerase is a (modified) host enzyme, the production of which is stimulated by the virus infection.

The insertion of "(modified)" is ambiguous, as are the multiple data based on annealing experiments to prove that sequences complementary to AlMV RNA are present in the products. These authors do not appreciate the significance of the fact that although the template–non-specific host enzyme synthesizes (often short) sequences complementary to the supplied template, the significance of such synthesis to viral replication *in vivo* is dubious (see Section V). This is a long and, to us, confusing paper that neglects some of the appropriate distinctions between enzymes from healthy and infected tissues made in the previous paper (Weening and Bol, 1975) from the same laboratory. A third paper (Linthorst *et al*, 1980) is yet more disconcerting in this respect since no attempt was made to characterize products synthesized by crude (membrane-bound) extracts from healthy, mock-inoculated, or tobacco leaves that were infected by wild-type or nine temperature sensitive AlMV mutants. No significant differences in radioactivity incorporated into acid-precipitable material were obtained for any of these extracts, and this potentially valuable series of experiments would be well worth repeating using purified extracts. The summary that 'it is very unlikely that a virus-specific protein is involved in these *in vitro* RNA polymerase activities" may be strictly correct, but is very misleading since it is easy to assume, incorrectly, that such a conclusion is warranted for the situation *in vivo*, or even for better defined experiments *in vitro*.

A lengthy paper by Le Roy *et al.* (1977) on polymerases in uninfected and AlMV-infected tobacco plants is of minimal value because of the very low activity reported. Almost all the data, including a table on template specificity, are based on reactions yielding less than 1000 cpm of acid-insoluble products. A subsequent paper from the same laboratory (Chifflot *et al.*, 1980) presents hybridization data as evidence for synthesis of new (+) strands. While the data do substantiate this claim, polyacrylamide gel electrophoresis of [^{3}H]RNA products synthesized in the presence of a DEAE-purified enzyme shows that the vast majority were no longer than single-stranded AlMV RNAs. Therefore, it is likely that any (+) strands were copied from (−) strand fragments. This casts great doubt as to whether the process represents the real viral replicative process. Further, the fact that the enzyme could be obtained from healthy or infected plants and exhibited no template specificity (a deficiency which we have detailed above) suggests that this enzyme is not the true viral replicase.

G. Turnip Yellow Mosaic Virus Replicase

Considerable success in isolating a polymerase fraction from TYMV-infected Chinese cabbage (*Brassica sinensis*) leaves has been achieved by Bové and his group at the Université de Bordeaux II. A series of papers (Bové *et al.*, 1965; 1968) led to the implication of the chloroplast membrane as the site of replication of TYMV RNA (Laflèche *et al.*,

1972). Bové *et al.* (1972) used the neutral detergent Lubrol to solubilize the replicase as an enzyme–template complex and Mouchès *et al.* (1974) freed the enzyme from its template by the use of a polyethylene glycol–dextran two phase system in the presence of 4M NaCl. In several respects the system adopted by Hardy *et al.* (1979) for BMV RNA polymerase resembles this system, but the use of an ammonium sulphate precipitation step for the TYMV enzyme may account for the generally lower activity of that enzyme because we (unpublished observations) have found that high salts inhibit BMV replicase. Mouchès *et al.* (1975) showed incorporation of [^{32}P]UMP (sp. act. 66 μCi/mmol) into acid-precipitable material from a background of 165 cpm to 3165 cpm by the addition of TYMV RNA as template. A considerable degree of template specificity was shown, BMV RNA and TMV RNA being 19% and 15% as effective templates as TYMV RNA. However, RNA from EMV (eggplant mosaic virus, a virus related to TYMV) was 70% as effective a template. The products of the reaction were shown to be double-stranded TYMV RNA by electrophoresis; however, the use of a gel-slicing technique available at that time rather than newer slab–gel separations complicates interpretation of the data. Demonstration of an association between the enzyme and TYMV RNA was performed using a nitrocellulose filter-binding technique. This approach was apparently later modified by Zabel *et al.* (1979) for their CPMV replicase assay.

Despite the intrinsic interest of the paper by Mouchès *et al.* (1975), there was little evidence as to the protein composition of the enzyme. The large size (400 000 daltons, estimated by sucrose density gradient centrifugation) led the authors to surmise that it comprises several subunits. They note that no TYMV RNA-specific activity is found in healthy leaves and that RNA polymerase activity found in such leaves (Astier-Manifacier and Cornuet, 1971) sedimented at 6S under conditions where the TYMV-dependent enzyme sedimented at 14S.

In a later paper (Garnier *et al.*, 1980), the nature of the RNA synthesized by TYMV replicase was further studied by high resolution autoradiography of glycol methacrylate-embedded ultrathin sections. The authors concluded that the (−) RNA template and nascent (+) RNA strands of the RI were essentially single-stranded, a stretch of only about five nucleotide pairs being hydrogen bonded by the replicase. This structure would be similar to that shown in Fig. 1–1.

H. Pea Enation Mosaic Virus Replicase

Like TYMV, PEMV appears to replicate in membrane-associated structures. However, the nuclear membrane, rather than the chloroplast membrane, is implicated (de Zoeten *et al.*, 1976; Powell and de Zoeten, 1977). Powell and de Zoeten (1977) reported that nuclei extracted from healthy pea (*Pisum sativum* L. *cv.* "Perfected Wales") leaves were "primed" by resuspension for about 1 h in a polymerase assay buffer. Subsequently, [^{3}H]UMP incorporation into acid-insoluble material was followed in reactions that contained several different viral RNA templates. PEMV RNA was ten times more effective as template than was TMV RNA (44·6 *vs* 4·5 pmol incorporated mg^{-1} protein). Hybridization to [^{125}I]-labelled PEMV RNA showed that some 17% of the nuclear product synthesized on addition of PEMV RNA to the reaction hybridized; no such hybridization was seen in products of reactions containing nuclei from healthy leaves that were not "primed" with PEMV.

The experiments of Powell and de Zoeten (1977) were interesting because they showed the ability of nuclear preparations to support viral replication. This organellar system provides information analogous to that obtainable from protoplast systems, but in its present stage of development, provides little additional information about the enzymology of replication of viral RNA.

I. Cucumber Mosaic Virus Replicase

Symons and co-workers have undertaken a number of investigations into the replication of CMV RNA (May *et al.*, 1969; 1970; May and Symons, 1971). Kumarasamy and Symons (1979) noted that extracts from CMV-infected cucumber seedlings used poly(C) and poly(U) as efficient templates for [^{32}p]UMP incorporation whereas poly(A) and poly(G) were not copied. The ability of $Q\beta$ replicase to copy poly(C) efficiently was used by Carmichael (1975) as the basis for RNA homopolymer chromatography of $Q\beta$ replicase. Similarly, Kumarasamy and Symons (1979) added purification on poly(C)-cellulose to other chromatographic steps (on phosphocellulose and DEAE cellulose) following an initial low salt polyethylene glycol precipitation of CMV RNA polymerase from a crude extract.

The enzyme obtained by poly(C)-cellulose chromatography was purified over 10 000-fold on the basis of [^{3}H]UMP incorporated mg^{-1} protein in the enzyme fraction. Electrophoretic separation of extracts from healthy and CMV-infected tissues clearly showed that a polypeptide of about 100 000 daltons was present only in the extracts from infected tissue. *In vitro* translation of CMV RNAs 1 and 2 yields polypeptides of up to 105 000 and 120 000 daltons, respectively (Schwinghamer and Symons, 1977), and it is reasonable to believe that the polypeptide seen only in extracts of CMV-infected tissues is one of these virally-coded proteins.

Despite the high specific activity of the CMV replicase isolated by Kumarasamy and Symons (1979), it showed little discrimination against other viral RNAs. BMV RNA, AlMV RNA, and TMV RNA were nearly as good, or better templates than was CMV RNA. Only short polynucleotide products (about 150 residues) were obtained. Nevertheless, these authors stress that no replicase activity was found at any stage when healthy plant material was taken through the same purification procedure that led to high activity for extracts of CMV-infected tissue. They believe CMV RNA replicase to be multicomponent enzyme consisting of viral-coded and host-coded polypetides analogous to the bacteriophage Q*β* and f2 replicases (Kamen, 1975; Federoff, 1975).

X. FUTURE DIRECTIONS

The preceding is by no means a comprehensive review of all of the work undertaken on the isolation of plant viral RNA replicases, and we apologize to those authors whose work was omitted. We also acknowledge that the observations recorded are slanted to reflect our belief that most, or all, viral replicases contain a virally-encoded polypeptide that functions in template selection.

Most authors cited note that relatively little polymerase activity exists in uninfected, or at least in tissue not intentionally inoculated, with the virus being studied. Where products have been examined, viral-specific RNA is produced much more abundantly by extracts from virus-infected tissue than from uninfected tissue, and the length of apparently "homologous" products from uninfected tissue appears generally to be very short.

In all cases, the initial activity is associated in a "particulate" fraction, i.e. in association with membranes. Various methods for eluting replicase activity from membranes so as to "solubilize" the enzyme have been employed, and we have stressed that most of these yield activity that is sparingly, or not at all, template-specific.

In the most highly purified enzyme fractions described thus far, those for CMV replicase (Kumarasamy and Symons, 1979) and BMV replicase (Fig. 5), a polypeptide thought to be virally-coded can be detected. It is of importance that the co-identity of BMV RNA 1 translation product and the BMV replicase polypeptide has been established, but the critical demonstration that it is functional in the replicase reaction has yet to be established. Recently, inhibition of TYMV replicase activity by antibodies made to host elongation factor has been demonstrated (Mouchès *et al.*, 1981), supporting the notion that both host and viral proteins are present in the active replicase.

The replication of viral RNA genomes is, to all appearances, different from the synthesis of host RNA. Thus, it would seem possible that a detailed understanding of the biochemistry of viral RNA synthesis could lead to the development of procedures that would block viral RNA replication without harming host functions. Elucidation of the structure of RNA-dependent RNA polymerase is crucial to such understanding, and it is to be hoped that this review will help stimulate interest in solving this important and vexing problem.

ACKNOWLEDGEMENTS

Aspects of research presented in this review were supported by grants to the University of Wisconsin from N.I.H. (AI 11572) and N.S.F. (PCM 78–22526).

REFERENCES

Ahlquist, P., Dasgupta, R. and Kaesberg, P. (1981). *Cell* **23,** 183–189.

Assink, J. M. H., Swaans, H. and van Kammen, A. (1973). *Virology* **53,** 384–391.

Astier-Manifacier, S. and Cornuet, P. (1971). *Biochim. Biophys. Acta* **232,** 484–493.

Atkins, J. F., Steitz, J. A., Anderson, C. W. and Model, P. (1979). *Cell* **18,** 247–256.

Baltimore, D and Franklin, R. M. (1962). *Biochem. Biophys. Res. Commun.* **9,** 388–392.

Barrell, B. G., Air, G. M. and Hutchinson III, C. A. (1976). *Nature* **264,** 34–41.

Becker, W. M. (1979). *In* "Nucleic Acids in Plants", (T. C. Hall and J. W. Davies, eds) Vol. 1, pp. 111–141. CRC Press, Boca Raton, Fla.

Beremand, M. and Blumenthal, T. (1979). *Cell* **18,** 257–266.

Blumenthal, T. and Carmichael, G. G. (1979). *Ann. Rev. Biochem.* **48,** 525–548.

Blumenthal, T. and Landers, T. A. (1976). *Biochemistry* **15,** 422–425.

Boege, F. and Sänger, H. L. (1980). *FEBS Lett.* **121,** 91–96.

Bové, J. M., Bové, C. and Mocquot, B. (1968). *Biochem. Biophys. Res. Commun.* **32,** 480–486.

Bové, J. M., Bové, C., Rondot, M. J. and Morel, G. (1965). *In* "Biochemistry of Chloroplasts", Vol. 2, Academic Press, London and New York, 329–339.

Bové, C., Mocquot, B. and Bové, J. M. (1972). *Symp. Biol. Hung.* **13,** "Nucleic

Acids and Proteins in Higher Plants", (G. L. Farkas, ed.), 43–59, Akademiai Kiado, Budapest.

Bradley, D. W. and Zaitlin, M. (1971). *Virology* **45,** 192–199.

Brinshammar, S. and Juntti, N. (1974). *Virology* **59,** 245–253.

Bujarski, J., Hardy, S. F., Miller, W. A. and Hall, T. C. (1982). *Virology* (in press).

Carmichael, G. G., Weber, K., Niveleau, A. and Wahba, A. (1975). *J. Biol. Chem.* **250,** 3607–3612.

Chambon, P. (1975). *Ann. Rev. Biochem.* **44,** 613–638.

Chifflot, S., Sommer, P., Hartmann, D., Stussi-Garaud, C. and Hirth, L. (1980). *Virology* **100,** 91–100.

Clerx, C. M. and Bol, J. F. (1978). *Virology* **86,** 241–253.

Davies, J. W. (1979). *In* "Nucleic Acids in Plants", (T. C. Hall and J. W. Davies, eds) Vol. II, pp. 111–149. CRC Press, Boca Raton, Fla.

de Zoeten, G. A., Powell, C. A., Gaard, G. and German, T. L. (1976). *Virology* **70,** 459–469.

de Zoeten, G. A., Assink, J. M. H. and van Kammen, A. (1974). *Virology* **59,** 341–355.

Dorssers, L., Zabel, P., Van Der Meer, J. and Van Kammen, A. (1982). *Virology* **116,** 236–249.

Duda, C. T. (1979). *Virology* **92,** 180–189.

Duda, C. T., Zaitlin, M. and Siegel, A. (1973). *Biochim. Biophys. Acta* **319,** 62–71.

Federoff, N. (1975). *Biochim. Biophys. Acta* **319,** 235–258.

Flanegan, J. B., Pettersson, R. F., Ambros, V., Hewlett, M. J. and Baltimore, D. (1977). *Proc. Natl. Acad. Sci. USA* **74,** 961–965.

Fraenkel-Conrat, H. (1979). *Trends in Biochemical Sciences*, August, 184–186.

Goldbach, R., Rezelman, G. and Van Kammen, A. (1980). *Nature* **286,** 297–300.

Garnier, M., Mamoun, R. and Bové, J. M. (1980). *Virology* **104,** 357–374.

Hadidi, A. and Fraenkel-Conrat, H. (1973). *Virology* **52,** 363–372.

Hardy, S. F., German, T. L., Loesch-Fries, L. S. and Hall, T. C. (1979). *Proc. Natl. Acad. Sci. USA* **76,** 4956–4960.

Heisenberg, M. and Schramm, G. (1965). *Z. Naturforsch.* **B20,** 454–457.

Ikegami, M. and Fraenkel-Conrat, H. (1978a). *Proc. Natl. Acad. Sci. USA* **75,** 2122–2124.

Ikegami, M. and Fraenkel-Conrat, H. (1978b). *FEBS Lett.* **96,** 197–200.

Ikegami, M. and Fraenkel-Conrat, H. (1979). *J. Biol. Chem.* **254,** 149–154.

Ikegami, M. and Fraenkel-Conrat, H. (1980). *Virology* **100,** 185–188.

Kamen, R. I. (1975) *In* "RNA Phages", (N. Zinder, ed.), pp. 203–234. Cold Spring Harbor Laboratory, Cold Spring Harbor, NY.

Kiberstis, P., Loesch-Fries, L. S., and Hall, T. C. (1981). *Virology* **112,** 804–808.

Kumarasamy, R. and Symons, R. H. (1979). *Virology* **96,** 622–632.

Kummert, J. and Semal, J. (1974). *Virology* **60,** 390–397.

Kummert, J. and Semal, J. (1977). *Virology* **77,** 212–220.

Laflèche, D., Bové, C., Dupont, G., Mouchès, C., Astier, T., Garnier, M. and

Bové, J. M. (1972). Proc. 8th FEBS Meeting, Amsterdam. RNA viruses/ribosomes **27,** 43–71. North Holland Publ., Amsterdam.

LeRoy, C., Stussi-Garaud, C. and Hirth, L. (1977). *Virology* **82,** 48–62.

Lindhorst, H. J. M., Bol, J. F. and Jaspars, E. M. J. (1980). *J. Gen. Virol.* **46,** 511–515.

Loesch-Fries, L. S. and Hall, T. C. (1980). *J. Gen. Virol.* **47,** 323–332.

Matthews, R. E. F. (1978). *Nature* **276,** 850.

May, J. T., Gilliland, J. M. and Symons, R. H. (1969). *Virology* **39,** 54–65.

May, J. T., Gilliland, J. M. and Symons, R. H. (1970). *Virology* **41,** 653–664.

May, J. T. and Symons, R. H. (1971). *Virology* **44,** 517–526.

Mouchès, C., Bové, C., Barreau, C. and Bové, J. M. (1975). INSERM (Les Colloques de l'Institut National de la Santé et de la Recherche Médicale) **47,** 109–120.

Mouchès, C., Bové, C. and Bové, J. M. (1974). *Virology* **58,** 409–423.

Mouchès, C., Renaudin, J., Chuchana, P. and Bové, J. M. (1981). 5th Int. Congress of Virology, Strasbourg, Abstracts, p. 249.

Pallansch, M. A., Kew, O. M., Palmenberg, A. C., Golini, F., Wimmer, E. and Rueckert, R. R. (1980). *J. Virol.* **35,** 414–419.

Palmenberg, A. and Kaesberg, P. (1974). *Proc. Natl. Acad. Sci. USA* **71,** 1371–1375.

Palmenberg, A. C., Pallansch, M. A. and Rueckert, R. (1979). *J. Virol.* **32,** 770–788.

Pelham, H. R. B. (1978). *Eur. J. Biochem.* **85,** 457–462.

Pelham, H. R. and Jackson, R. J. (1976). *Eur. J. Biochem.* **67,** 247–256.

Powell, C. A. and de Zoeten, G. A. (1977). *Proc. Natl. Acad. Sci. USA* **74,** 2919–2922.

Powell, C. A., de Zoeten, G. A. and Gaard, G. (1977). *Virology* **78,** 135–143.

Rackwitz, H. R., Rohde, W. and Sänger, H. L. (1981). *Nature* **291,** 297–301.

Ralph, R. K. and Wojcik, S. J. (1969). *Virology* **37,** 276–282.

Röder, A. and Koschel, K. (1975). *J. Gen. Virol.* **28,** 85–98.

Roeder, R. G. (1976). *In* "RNA Polymerase", (R. Losick and M. Chamberlin, eds), pp. 285–329. Cold Spring Harbor Laboratory, Cold Spring Harbor, NY.

Romaine, C. P. and Zaitlin, M. (1978). *Virology* **86,** 241–253.

Rueckert, R. R., Palmenberg, A. C. and Pallansch, M. A. (1980). *Eur. J. Cell Biol.* **22,** 263–275.

Schwinghamer, M. W. and Symons, R. H. (1977). *Virology* **79,** 88–108.

Sela, I. and Hauschner, A. (1975). *Virology* **64,** 284–288.

Semal, J. and Hamilton, R. I. (1968). *Virology* **36,** 293–297.

Semal, J. and Kummert, J. (1970). *J. Gen. Virol.* **7,** 173–176.

Semal, J. and Kummert, J. (1971). *J. Gen. Virol.* **10,** 79–89.

Semal, J. and Kummert, J. (1976). *Ann. Microbiol.* (Inst. Pasteur) 127A, 153–158.

Shih, D. S., Shih, C. T., Kew, O., Pallansch, M., Rueckert, R. and Kaesberg, P. (1978). *Proc. Natl. Acad. Sci. USA* **75,** 5807–5811.

Shih, D. S., Shih, C. T., Zimmern, D., Rueckert, R. R. and Kaesberg, P. (1979). *J. Virol.* **30,** 472–480.

Southern, E. M. (1975). *J. Mol. Biol.* **98,** 503–517.

Symons, R. H. (1977). *Nucl. Acids Res.* **4,** 4347–4355.

Temin, H. and Baltimore, D. (1972). "Advances in Virus Research", (K. M. Smith and M. A. Lauffer, eds), Vol. 17, pp. 129–186, Academic Press, New York and London.

Traub, A., Diskin, B., Rosenberg, H. and Kalmar, E. (1976). *J. Virol.* **18,** 375–382.

Van Tol, A. and Van Vloten-Doting (1979). *Eur. J. Biochem.* **93,** 461–466.

Van Vloten-Doting, L., Neeleman, L. and Van Tol, R. G. L. (1980). 6th EMBO Annual Symposium 1980 "Molecular Biologists Look at Green Plants", Sept. 29–Oct. 2, EMBL Heidelberg, Proceedings, pp. 59–60.

Weening, G. J. and Bol, J. F. (1975). *Virology* **63,** 77–83.

Weinand, U., Bruschke, C. and Feix, G. (1979). *Nucl. Acid. Res.* **6,** 2707–2715.

Weissman, C., Simon, L. and Ochoa, S. (1963). *Proc. Natl. Acad. Sci. USA* **49,** 407–414.

White, J. L. and Dawson, W. O. (1978). *Virology* **88,** 33–43.

White, J. L. and Murakishi, H. H. (1977). *J. Virol.* **21,** 484–492.

Zabel, P. (1978). Thesis, Wageningen, p. 88.

Zabel, P., Dorssers, L., Wernars, K. and Van Kammen, A. (1981). *Nucl. Acid. Res* **9,** 2433–2453.

Zabel, P. Jongen-Neven, I. and Van Kammen, A. (1976). *J. Virol.* **17,** 679–685.

Zabel, P., Jongen-Neven, I. and Van Kammen, A. (1979). *J. Virol.* **29,** 21–33.

Zabel, P., Weenen-Swaans, H. and Van Kammen, A. (1974). *J. Virol.* **14,** 1049–1055.

Index

A

Active defence, *see* Resistance
Alternaria alternata toxin, *see* AA1 toxin
AA1 toxin
  inhibition of aspartate carbamyl transferase, 118
  structure, 102–104
Agrobacterium rhizogones, 153
Agrobacterium radiobacter, 153, 158–159
Agrobacterium rubi, 153–154
Agrobacterium tumefaciens, *see also* Crown gall and Ti plasmid
  as cause of crown gall, 140–141
  biology of infection process, 147–150.
  classification based on opine metabolism, 154–157
  expression of T-DNA, 143–144
  model for microbial evolution, 170
  opine utilization, 144–145
  plasmid, 141–142
  taxonomy, 153–154
  transfer of plasmid DNA to host, 142–143
Agrocin-84, 158–159
Agropene, 155–157, 160
AK toxin, structure, 98
Alfalfa mosaic virus, 180–208 *passim*
  replicase, 203–204
Alternaria kikuchiana toxin, *see* AK toxin
Alternaria mali toxins, *see* AM toxins
Alternaria solani, toxin structure, 105
AM toxins, structure, 98–99
AMV, *see* Alfalfa mosaic virus
Anastamosis, role in cytoplasmic hypovirulence, 19, 22–24
Animal viral replicase, 189
Auxin, production in Crown gall, 159–160

B

Bacteria
  gene-for-gene systems, 51–52
  surface molecules, 58–59
  viral replicase, 188–189
Barley
  genetics of resistance, 44–53
  horizontal resistance, 50
  induced susceptibility, 41–42
  viral replicases, 180–208 *passim*
BBMV, *see* Broad bean mosaic virus
Biological control
  crown gall, 158–159
  Endothia parasitica, 12
BMV, *see* Brome mosaic virus
Broad bean mosaic virus, 180–208 *passim*
Brome mosaic virus, 180–208 *passim*
  isolation of replicase, 193–196
  properties of replicase, 187–188
  structure of replicase, 187–188

C

Cancer, 170
Castanea crenata, as host of *E. parasitica* in Asia, 12
Castanea dentata
  as host of *E. parasitica* in North America, 3–7
  natural range of, 5–6
Castanea mollisima, as host of *E. parasitica* in Asia, 12

Castanea pumilla
as host of *E. parasitica* in Asia, 12
virulence of *E. parasitica* to, 5–6
Castanea sativa, as host of *E. parasitica* in Europe, 7–12
CCMV, see cowpea chlorotic mottle virus
Ceratocystis ulmi, hypovirulence in, 28–29
Chestnut blight
Asia, 12
Europe, 7–12
North America, 3–7
Chloroplasts, effects of T toxin, *see* T toxin
CMV, *see* Cucumber mosaic virus
Compatibility
basic (general), 36
specific, 37–39
Continental drift, role in origin of cytoplasmic hypovirulence, 20–22
Cowpea chlorotic mottle virus, 180–208 *passim*
replicase, 197
Cowpea mosaic virus, 180–208 *passim*
replicase, 197–200
CPMV, *see* Cowpea mosaic virus
Cross protection, 56
Crown gall, *see also Agrobacterium tumefaciens* and Ti-plasmid
biological control, 158–159
concept of genetic parasitism, 145–147
factors contributing to elucidation of mechanism, 172–174
model for growth regulation research, 169–170
model for host-pathogen research, 169
opine production, 144–145
persistence of transformed state, 141
prokaryotic gene expression in eukaryotic cell, 171
proof of bacterial etiology, 140–141
proof of Ti plasmid control of traits, 164–165
relation to animal cancers, 170
role of Ti plasmid, 141–142, 159–164
transfer of plasmid DNA, 142–143
variation in tumour form, 151–152
the infection process
conditioning phase, 148–149
developmental phase, 150–153
induction, 149–150
Cucumber mosaic virus, 180–208 *passim*
replicase, 206–207
Cytokinin, production in Crown gall, 159–160

E

Education, of plant pathologists, 172–174
Elicitor–receptor model, 62–71
allelic genes, 71
DNA transcription, 65–66
fast and slow genes, 69–70
functional basis for stabilizing selection, 69
fungal surface carbohydrates, 72–74
ghost genes, 70
mesothetic reactions, 70
phytoalexin systemicity, 66
rationalization of genetic inconsistencies, 67–68
role of glycoproteins, 64
role of glycosyl transferase, 68–71
second messenger, 65–66
soluble elicitors, 64–65
specific receptors, 71–72
EMCV, *see* Encephalomyocarditis virus
Encephalomyocarditis virus, 189–191
Endothia parasitica, *see also* Chestnut

blight and Hypovirulence
- accession to dominance of CH strains, 17–20
- abnormal cankers, 8–10, 16–17
- abnormal isolates, 10–11
- anastamosis, 19, 22–24
- cankers of, 4–5
- co-existence of CH strains, 15–16
- cytoplasmic hypovirulence in North America, 12–13
- debilitated CH strains, 17
- development of cytoplasmic incompatibility, 19–20
- diversity of CH strains, 15–16
- heterokaryosis, 22–24
- hybridization, 22–24
- instability of CH strains, 16
- interaction with *E. radicalis*, 24–27
- natural biological control, 15–20
- natural spread of CH strains, 17–19
- origin of cytoplasmic incompatibility, 20–28
- probable causes of hypovirulence, 13–15
- role of vectors in spread of CH strains, 18
- vegetative incompatibility, 19–20

Endothia radicalis, role in cytoplasmic hypovirulence in *E. parasitica*, 24–27

Erwinia amylovora, complex pathogenesis, 36

Erysiphe graminis
- genetics of virulence-avirulence, 44–53
- hypersensitivity to, 41–42
- induced susceptibility to, 41–42

Evolution, microbial, 170

Extracellular polysaccharides, of bacteria, 59

F

Fast genes, *see* Gene-for-gene systems

FC toxin
- interaction with hormones, 122–123
- mode of action, 119

Flax
- genetics of resistance, 44–53
- hypersensitive reaction, 41

Functional proteins, of viruses, 180–181

Fungi, surface molecules of, 59–60

Fusicoccin, see FC toxin

G

Gene dosage, *see* Gene-for-gene systems

Gene-for-gene systems, *see also* Elicitor–receptor model and Resistance
- allelism, 53, 71
- associated genetic factors, 49
- bacteria, 51–52
- complementarity, 44–47
- cross protection, 56
- elicitor–receptor model, 62–71
- *Erysiphe graminis tritici*, 46–47
- evolution, 46
- fast genes, 48, 69–70
- flax, 47
- fungal surface carbohydrates, 61–62, 72–73
- gene dosage, 48
- ghost genes, 46–47, 70
- horizontal resistance, 49–50
- host-selective toxins, 125–127
- linked genes, 50
- *Melampsora lini*, 47
- mixed infection types, 51
- phytoalexins, 53–56
- plant age, 50
- quadratic check, 44–45
- recessive resistance genes, 48
- reciprocal check, 44–45
- recognition mechanisms, 47–61, 56–57

Gene-for-gene systems *(Cont.)*
slow genes, 48
stabilizing selection, 51, 69
strong genes, 51
temperature effects, 50, 55–56
viruses, 51–52
weak genes, 51
wheat, 46–47
Genetic engineering, use of Ti plasmid as vector, 171–172
Genetic parasitism, concept of, 145–147
Ghost genes, *see* Gene-for-gene systems
Glyceollin, role in resistance, 56, 61, 65–66
Glycoproteins
role in active defence, 60–62
role in elicitor–receptor model, 64–71 *passim*
Glycosyl transferase, *see* Elicitor–receptor model

H

HC toxin
effects on the plasmalemma, 113
receptor sites, 123
structure, 95
Helminthosporium carbonum toxin, see HC toxin
Helminthosporium maydis
economic importance, 87
toxin, *see* T toxin
Helminthosporium sacchari toxin, *see* HS toxin
Helminthosporium victoriae
economic importance, 87
induced susceptibility to, 42
toxin, *see* HV toxin
Helminthosporoside, *see* HS toxin
Heterokaryosis, role in cytoplasmic hypovirulence, 22–24
Histopene, 155
Homooctopene, 155
Horizontal resistance, 49–50
HS toxin
binding to host protein, 117–118, 129–131
rate of production, 92–93
role in disease, 128–129
structure, 96–98
HV toxin
effects on plasmalemma, 111–113, 120–121
receptor sites, 123
role in disease, 128
self-repair hypothesis, 124–125
structure, 94–95
Hybridization, role in cytoplasmic hypovirulence, 22–24
Hypersensitive reaction
flax, 41
genetic basis, 40–41
induction by *Erysiphe graminis*, 40–41
induction by *Melampsora lini*, 41
relationship to animal immune systems, 43–44
role in active defence, 39–42
role of phytoalexins, 42–43
wheat, 40–41
Hypovirulence, *see also* Chestnut blight and *Endothia parasitica*
concept of, 2–3
cytoplasmic, 3
nuclear, 3
in *Ceratocystis ulmi*, 28–29
in *Rhizoctonia solani*, 29–30
role in *Endothia-Castanea* pathosystem
abnormal cankers as reservoirs of, 16
accession to dominance of CH strains, 17–20
anastamosis, 19, 22–24
co-existence of CH strains 15–16
debilitated CH strains, 17
discovery, 10–12
diversity of CH strains in North America, 15–16

effect of continental drift, 20–22
heterokaryosis, 22–24
hybridization, 22–24
instability, 16
natural spread of CH strains, 17–19
occurrence in North America, 12–13
origin, 20–28
probable causes of, 13–15
recovery of chestnuts in Italy, 12
role of ds RNA, 13–15
role of *Endothia radicalis*, 24–27

I

Immune system, in animals, 43–44
Immunoglobulin, 43–44
Indole-3-yl acetic acid, interaction with toxins, 122–123, *see also* Auxin
Induced susceptibility
role of host-specific toxins, 42
to *Erysiphe graminis hordei*, 41–42
Insects
gene-for-gene systems, 52
resistance to, 35–74 *passim*

L

Lectins, 59–60
Linked genes, in gene-for-gene systems, 50
Lysopine, 155

M

Major histocompatibility complex, 43–44
Melampsora lini
genetics of virulence-avirulence, 44–53
hypersensitive reaction to, 41
Mitochondria, effects of T toxin, 114–116
Mycoplasmas, surface molecules of, 57–58
Mycorrhiza, lack of resistance to, 38

N

Nematodes
gene-for-gene systems, 52
surface molecules, 57–58
Nopaline, 155–157, 160, 163
Nopalinic acid, 155

O

Obligate parasites, complex compatibility of, 38
Octopine, 155–157, 160, 161
Octopinic acid, 155
Opines, *see also* individual opines
use in classification of Agrobacteria, 154–157
production, 144–145
Ornaline, *see* Nopalinic acid
Overlapping genes, 180

P

Pathogenesis
complex 36–39
requirements for, 36–39
simple, 36
Pathotoxins, *see also* individual toxins
host-selective
bioassays, 90–92
definition of terms, 88–89
determination of structure, 93–106
distribution in taxonomic groups, 86–87
economic importance, 87–88
effects on chloroplasts, 116–117

Pathotoxins *(Cont.)*
effects on host plasmalemma, 111–114
effects on mitochondria, 114–116
effects on non-hosts, 109–110
genetic aspects, 125–127
historical background, 84
interaction with specific host compounds, 117–118
modes of action, 106–118
physical techniques for determining modes of action, 108–109
primary effects, 106–108
production in culture, 89–90
recognition sites, 123–124
relationship to hormones, 122–123
role of energy-transducing membranes, 119–121
role in induced susceptibility, 42
roles in pathogenesis, 127–129
scientific significance, 84–87
secondary effects, 106–108
similarity of modes of action, 119
host-specific, *see* Pathotoxins, host-selective
PC toxin
effects on plasmalemma, 113, 120–121
receptor sites, 123
structure, 95
Pea enation mosaic virus, 180–208 *passim*
replicase, 205–206
PEMV, *see* Pea enation mosaic virus
Periconia circinata toxin, *see* PC toxin
Phylosticta maydis, toxin structure, 104
Phytoalexins
association with the hypersensitive reaction, 42–43
degradative enzymes, 43, 54
effects on pathogens, 53
elicitor–receptor model, 62–71
mutations, 74
rate of accumulation, 54–55
role in gene-for-gene systems, 53–56
systemicity, 66
Phytophthora megasperma
elicitor–receptor model, 64–66, 73
surface carbohydrates, 61–62
Plasmalemma
effects of toxins on, 111–114
Plasmid, role in crown gall, *see* Ti plasmid
Polymerase, RNA, *see also* Replicase
activity in healthy plants, 184–185
definition, 181–184
model of replication complex, 187
Proteins, viral
interaction with host protein and nucleic acid, 181
processing of, 181
Pseudomonas solanacearum, complex pathogenesis of, 36
Puccinia graminis, genetics of virulence-avirulence, 44–53
Puccinia recondita, genetics of virulence-avirulence, 44–53

Q

Quadratic check, *see* Gene-for-gene systems

R

Receptors, for host-selective toxins, 123–124
Reciprocal check, *see* Gene-for-gene systems
Recognition mechanisms
animal systems, 43–44
fungal surface carbohydrates, 61–62, 72–74
gene-for-gene systems, 56–57
genetic constraints, 47–53
specific receptors of plants, 71–72

surface molecules of pathogens, 57–60, 72–74
Replicase, viral
alfalfa mosaic virus, 203–204
animal viruses, 189
bacterial viruses, 188–189
brome mosaic virus, 192–196
cowpea chlorotic mottle virus, 197
cowpea mosaic virus, 197–200
cucumber mosaic virus, 206–207
definition, 181–184
evidence for host encoding, 185–186
existence of, 185–187
isolation from plants, 189–192
detrimental effect of high ionic conditions, 191–192
ethanol as a poison, 191
evidence for existence of processing step(s), 190–191
membrane impurities, 191–192
use of detergents, 192
pea enation mosaic virus, 205–206
research needs, 207–208
structure in plants, 187–188
turnip yellow mosaic virus, 204–205
Resistance, *see also* Elicitor–receptor model and Gene for-gene systems
elicitor–receptor model, 62–71
genetic basis 44–53
hypersensitivity, 39–42
mechanisms, 38–75
non-host, 38
recognition mechanisms, 56–57
role of phytoalexins, 53–56
specific, 38
Rhizobium spp., lack of resistance to, 38–39
Rhizoctonia solani, cytoplasmic hypovirulence in, 29–30
Rhyncosporium secalis, toxin structure, 104–105
RNA, ds
in *Ceratocystis ulmi*, 29
in *Endothia parasitica*, 13–15
in *Rhizoctonia solani*, 30
RNA polymerase, *see* Polymerase and Replicase

S

Satellite tobacco necrosis virus, 180–208 *passim*
Sclerotium rolfsii, simple pathogenesis of, 36
Slow genes, *see* Gene-for-gene systems
Stabilizing selection, *see* Gene-for-gene systems
STNV, *see* Satellite tobacco necrosis virus
Strong genes, *see* Gene-for-gene systems

T

T-DNA, *see* Ti plasmid
Teratomas, 151–152
TIP, in crown gall, 141–142
Ti plasmid, *also see Agrobacterium tumefaciens* and Crown gall
attachment of T-DNA to host DNA, 162–164
common DNA, 167
expression of T-DNA in plants, 143–144, 150–151
functional organization of nopaline type, 163
functional organization of octopine type, 161
genes for opine synthesis and catabolism, 168
genetic engineering, 171–172
homology of octopine and nopaline types, 167–168
incompatibility between strains, 168
mutations, 168

Ti plasmid *(Cont.)*
  opine synthesis and catabolism, 144–145, 154–156
  physical and genetic maps, 165–168
  plasmid-determined characters, 166
  proof of control of traits, 164–165
  role in crown gall, 141–142
  role in tumour induction, 149–150, 159–164
  transfer to plant tumour cells, 142–143
TMV, *see* Tobacco mosaic virus
TNV, *see* Tobacco necrosis virus
Tobacco, resistance to TMV, 50
Tobacco mosaic virus, 180–208 *passim*
  evaluation of existence of replicase, 185
  replicase, 200–203
  resistance to, 50
Tobacco necrosis virus, 180–208 *passim*
Transformation, 141, 149–150
T-toxins
  bioassays, 90–92
  effects on chloroplasts, 116–117
  effects on energy-transducing membranes, 121
  effects on mitochondria, 114–116
  effects on the plasmalemma, 113–114, 120–121
  functioning as ionophores, 122
  interaction with hormones, 122–123
  receptor sites, 123–124
  role in disease, 128–129
  structure, 99–102
Turnip yellow mosaic virus, 180–208 *passim*
  replicase, 204–205
TYMV, *see* Turnip yellow mosaic virus

V

Vectors, role in spread of CH strains of *E. parasitica*, 18
Vegetative incompatibility, role in cytoplasmic hypovirulence, 19–20
Victorin, *see* HV toxin
Virulence, concept of, 2
Virulence genes, 44–47, 52–53
Viruses, *see also* individual viruses
  functional proteins, 180–181
  gene-for-gene systems, 51–52
  overlapping genes, 180
  replicase, 185–187, *see also* Replicase, viral
  resistance to, 35–74
  RNA replication pathways, 182–184
  surface molecules, 57–58
Virus-like agents, role in cytoplasmic hypovirulence, 14–15

W

Weak genes, *see* Gene-for-gene systems
Wheat
  genetics of resistance, 44–53
  hypersensitive reaction, 40–42